Platelet Rich Plasma in Medicine

Elie M. Ferneini

Regina Landesberg · Steven Halepas

Editors

Platelet Rich Plasma in Medicine

Basic Aspects and Clinical Applications

Editors
Elie M. Ferneini
Department of Surgery
Frank H Netter MD School of Medicine
Quinnipiac University
North Haven, CT, USA

Regina Landesberg
Simsbury, CT, USA

Steven Halepas
Columbia University Irving Medical Center
NewYork–Presbyterian Hospital
New York, NY, USA

ISBN 978-3-030-94268-7   ISBN 978-3-030-94269-4   (eBook)
https://doi.org/10.1007/978-3-030-94269-4

This Springer imprint is published by the registered company Springer Nature Switzerland AG
The registered company address is: Gewerbestrasse 11, 6330 Cham, Switzerland

Contents

Contributors

Lea Bach, MS Department of Orthopaedics, Rutgers New Jersey Medical School, Newark, NJ, USA

Xun Joy Chen, DMD, MD Will Surgical Arts, Rockville, MD, USA

Andrew R. Emery, DMD, MD Department of Oral and Maxillofacial Surgery, Massachusetts General Hospital/Harvard University, Boston, MA, USA

Elie M. Ferneini, DMD, MD, MHS, MBA, FACS, FACD Beau Visage Med Spa and Greater Waterbury OMS, Cheshire, CT, USA

Department of Surgery, Frank H Netter MD School of Medicine, Quinnipiac University, North Haven, CT, USA

Division of Oral and Maxillofacial Surgery, University of Connecticut, Farmington, CT, USA

Michael S. Forman, DMD, MD Division of Oral and Maxillofacial Surgery, NewYork-Presbyterian/Columbia University Irving Medical Center, New York, NY, USA

Nicholas Genovese, MD Department of Orthopaedics, Rutgers New Jersey Medical School, Newark, NJ, USA

Steven Halepas, DMD, MD, FACS Division of Oral and Maxillofacial Surgery, NewYork-Presbyterian/Columbia University Irving Medical Center, New York, NY, USA

Alia Koch, DDS, MD, FACS Division of Oral and Maxillofacial Surgery, NewYork-Presbyterian/Columbia University Irving Medical Center, New York, NY, USA

Regina Landesberg, DMD, PhD Private Practice, Oral and Maxillofacial Surgery, Torrington, CT, USA

Katherine Lauritsen, BS Department of Orthopaedics, Rutgers New Jersey Medical School, Newark, NJ, USA

Kevin C. Lee, DDS, MD Division of Oral and Maxillofacial Surgery, NewYork-Presbyterian/Columbia University Irving Medical Center, New York, NY, USA

Sheldon S. Lin, MD Department of Orthopaedics, Rutgers New Jersey Medical School, Newark, NJ, USA

Michael Metrione, MD Department of Orthopaedics, Rutgers New Jersey Medical School, Newark, NJ, USA

Keyur Naik, DDS, MD Department of Oral and Maxillofacial Surgery, New York University Langone Medical Center/Bellevue Hospital Center, New York, NY, USA

Lakshmi S. Nair, PhD Department of Orthopaedic Surgery, University of Connecticut Health Center, Farmington, CT, USA

The Connecticut Convergence Institute for Translation in Regenerative Engineering, University of Connecticut Health Center, Farmington, CT, USA

Department of Materials Science and Engineering, Institute of Material Science, University of Connecticut, Storrs, CT, USA

Department of Biomedical Engineering, University of Connecticut, Storrs, CT, USA

Alexander Pascal, DDS Division of Oral and Maxillofacial Surgery, New York-Presbyterian/Columbia University Irving Medical Center, New York, NY, USA

PRP History

Michael S. Forman and Alia Koch

Introduction

The clinical applications and use of platelet-rich therapy (PRT) in medicine and surgery have thrived over the past two decades. More specifically, platelet-rich plasma (PRP) is an autologous blood product which contains a concentration of platelets that is above the physiologic baseline for whole blood, reported to be three to five times the normal value [1, 2]. PRP harnesses the signaling molecules and growth factors of platelets such as vascular endothelial growth factor (VEGF), fibroblast growth factor (FGF), platelet-derived growth factor (PDGF), epidermal growth factor, insulin-like growth factors (e.g., IGF-1, IFF-2), matrix metalloproteinases (MMPs), and interleukin 8 (IL-8).

The concept was first described and developed in the field of hematology as early as 1951 as a method to treat thrombocytopenia [3–5]. Interestingly, it was the field of oral and maxillofacial surgery in the 1990s that really developed and repurposed this biology for regenerative techniques. Over the past 20 years, its application has spread significantly to other fields, particularly those dealing with musculoskeletal problems, orthopedics, and cosmetic procedures, as well as cardiac surgery and plastic surgery.

In order to appropriately understand the history of PRP and to recognize how far this field has advanced, it is important to first appreciate the scientific discovery of the importance of blood itself, its therapeutic potential, and innovative devices that enabled blood to be separated into its components.

M. S. Forman (✉) · A. Koch
Division of Oral and Maxillofacial Surgery, NewYork-Presbyterian/Columbia University
Irving Medical Center, New York, NY, USA
e-mail: Mf3097@cumc.columbia.edu

© The Author(s), under exclusive license to Springer Nature Switzerland AG 2022
E. M. Ferneini et al. (eds.), *Platelet Rich Plasma in Medicine*,
https://doi.org/10.1007/978-3-030-94269-4_1

Where PRP Started: History of Circulation and Blood Therapies

PRT would not be possible today if the earliest scientists did not first discover blood itself and its ability to be used as a medical therapy (see Fig. 1.1). The first observation that blood was critical to life can be traced back in the scientific literature as early as the seventeenth century. In 1615, a German physician, Andreas Libavius, first described the power of blood and its potential for transfusion by advocating for an artery-artery connection, writing "…[the] spirituous blood of the young man will pour into the old one as if it were from a fountain of life, and all of his weakness will be dispelled [6]."

The first known application of blood use for medical therapy was for blood transfusions. Generally credited with the creation of blood transfusions was an English Physician, William Harvey, who first published a description of blood circulation in *1628* [6]. Previously, scientists believed that blood was simply produced constantly by the nutrients obtained from food and never stored or circulated. It was Harvey who first described the idea of circulation in his text, *De Motu Cordis*, in 1628 as, "It has been shown…that blood by the beat of the ventricles flows through the lungs and heart and is pumped to the whole body… it returns from the periphery everywhere to the center, from smaller veins to the larger ones…blood moves around in a circle continuously and that the action or function of the heart is to accomplish this by pumping [7]." This breakthrough concept inspired the next generation of physicians, who believed in harnessing circulating blood for its therapeutic potential.

The idea of blood transfusions between living organisms was originally reported by another English physician, Richard Lower, who *actually* exhibited in Oxford the effect of, first, exsanguinating a medium-sized dog and then, second, immediately

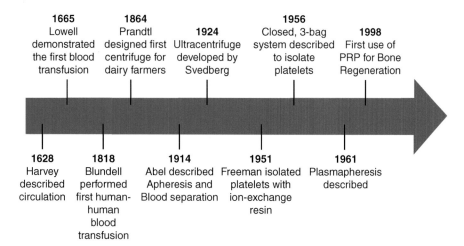

Fig. 1.1 A timeline of the notable achievements that led to the first use of PRP, including the discovery of blood (1628), centrifuge creation (1864), and the clinical use of PRP in present day (1998)

bringing it back to life after directly receiving donor blood from two large mastiffs in *1665* [6]. Just 2 years later in *1667*, multiple reports of physicians employing xenologous transfusions between animals (e.g., calves, lambs) and humans were published. However, this experimentation was eventually prohibited due to the severe adverse reactions that resulted.

It was not until *1818*, over 100 years later, when an English obstetrician, James Blundell, was widely credited for performing the first human-to-human blood transfusion [8, 9]. He reported a successful transfusion of his patient with her husband's blood for the treatment of postpartum hemorrhage, utilizing a syringe filled directly from the husband's vein which was then injected directly into his patient's vein [8]. Blundell continued to publish data on his experience, developed indications for transfusion, and innovated devices for transfusion which advanced the field. The next major breakthrough was in 1908, when Austrian physician, Karl Landsteiner, discovered A, B, and O blood groups for which he received the Nobel Prize in 1930.

Naturally, as the science and clinical therapeutic breakthroughs of transfusion medicine continued to flourish, it led to the novel idea of using a patient's own blood for its components. When did this start to happen? This generated an array of novel medical and surgical therapies including PRP.

Innovation and Device Discovery

While younger clinicians may take readily available and affordable centrifuges for granted in the twenty-first century, the technical difficulties of blood separation significantly hindered physicians just 100 years ago. A centrifuge is a critical device capable of separating particles of different size, shape, and density from solutions. It has a rapidly rotating container that applies centrifugal force to its contents. When blood is centrifuged, it is separated into plasma and erythrocytes, due to differences in their specific gravities. The platelets remain in the plasma, and hence PRP would not be possible today without an accessible and reliable device to separate the components of blood.

Surprisingly, it was a German engineer, Antonin Prandtl, who developed the concept of a centrifuge in 1864 for dairy farmers to separate cream from milk. Soon thereafter, Swedish Engineer, Karl Gustaf Patrik De Laval, filed a cream separator patent in 1878 for the first open, continuous-flow centrifuge which revolutionized the dairy industry and later the entire medical field [10].

In 1924, a Swedish chemist, Theodor Svedberg, developed the ultracentrifuge, which allowed for the determination of molecular weights of compounds [11]. With forces capable of reaching >100,000 g, this development enabled scientists to separate particles, determine weights, and conduct real-time analytics and photography during its use. The original ultracentrifuge was based off a rotor that was powered by a small turbine and pressurized oil, for which Svedberg earned a Nobel Prize in 1926 [12]. However, it was costly and difficult to manufacture, which limited its widespread adoption. In 1925, Jesse Beams, an American Physicist, developed an air turbine centrifuge that was powered by compressed air, and the speed was determined by the air pressure driving it [13]. This product was easier to produce, and by placing the air turbine in a vacuum chamber, it reduced friction, heating, and convection of the substance being centrifuged [12]. A student of Beams, Edward Pickels, eventually prepared the first commercially available centrifuge, the Model E, in the late 1940s, through a company called Spinco. As these centrifuges became more available, the devices ultimately allowed for separation of blood, cells, and molecules and led to major scientific discoveries: proteins were discrete molecular entities of defined size [11], DNA replicated in a semi-conservative nature [14], and hydrogen bonds existed between base pairs in DNA [15], among many more.

Blood Separation

This technology was immediately put to use for blood separation. The original concept of *separating* the portions of blood into plasma, leukocytes, erythrocytes, or platelets and then returning the desired portions to the patient as a therapy dates back to 1914, when plasmapheresis was first discovered [16]. The investigators collected large volumes of blood from dogs, used a centrifuge to discard the plasma (with platelets), and resuspended the erythrocytes in a more dilute solution before returning it [16]. The investigators used dogs to demonstrate the ability to safely withdraw large quantities of blood plasma as long as erythrocytes were resuspended and returned to the host [16].

Initial use for this was targeted to patients with hematologic malignancies, predominantly leukemias, to collect large volumes of leukocytes and return the remaining blood, in an efficient and continuous manner [17]. One of the original methods for the separation of leukocytes from whole blood was in 1947. Vallee et al. used differences in specific gravities of leukocytes and erythrocytes, along with albumin to centrifuge whole blood and collect leukocytes at the plasma-albumin interface [18, 19]. They layered whole blood on albumin solutions of various densities to identify one that was in between the density of erythrocytes and leukocytes, so that on centrifugation, red cells propelled through the albumin but leukocytes did not and settled at the plasma albumin interphase [18].

At the time of the Second World War, a large need developed for the storage of human blood components for transfusion. This need provided an opportunity for a group from Harvard, led by professor Dr. J. Edwin Cohn, to employ the concept of De Laval's cream separator to separate plasma from whole blood [10, 19]. In 1951, he developed the *Cohn Fractionator* which was able to successfully separate the components of blood into plasma, leukocytes, platelets, and erythrocytes in a continuous manner [20]. Although his device was an incredible engineering achievement and provided large volumes, it was costly, complex, and never succeeded commercially. How was this device so different than prior?

Large-scale and high-volume devices continued to be developed through the 1960s and 1970s and led to some novel achievements. These included the design of the first closed continuous-flow centrifuge via a collaboration between International Business Machines Corp. (IBM) and the National Cancer Institute (NCI) for treatment of hematologic malignancies [10, 21]. The continued advances in larger scale device innovation is beyond the scope of this chapter but helped pave the way for the work we do today related to PRP.

Platelet Separation and Isolation

As technology advanced with centrifuges and blood separation devices, the term "platelet-rich plasma" appeared in the literature during 1950s. However, in early investigations, PRP was merely mentioned when describing it as a clinically irrelevant first by-product of blood centrifugation [3]. In that context, red blood cells or white blood cells were the desired product. It was the oncology field that drove the research into efficient blood separation, specifically for leukocytes and platelets. The two most important causes of death in acute leukemia were massive hemorrhage secondary to thrombocytopenia and infection associated with leukopenia due to the replacement of the bone marrow with leukemic cells [4]. Hence there was a focus on improving survival and preventing hemorrhage in patients during active treatment, which would drive platelet counts dangerously low (<60,000 platelets/mm^3).

In 1951, Gustave Freeman, an American physician-scientist, was the first to describe a technique to isolate platelets that was discovered incidentally. While working with an ion-exchange resin to make blood uncoagulable by fixing calcium ions, he noted "…it proved, fortuitously to cause partial disappearance of platelets from the blood [5]." By passing 500 mL of donor blood through the resin column, and subsequently washing the columns with saline to recapture the platelets, he was able to obtain on average of 119,000 platelets/mm^3, or 40% of normal platelet levels [5]. For greater yield, he completed multiple passes of blood through four identical resin columns which increased the yield to 90–95% [5]. At the time he believed this method best maintained platelet morphological and physiological integrity that was superior to centrifuge and washing, which required multiple handlings and manipulations of the blood. What did the resin do to pull the platelets out?

In the following years, 1954–1956, a closed plastic system involving three separate but connected bags was described for isolating platelets [3] (Fig. 1.2). The apparatus was connected via tubing to the needle used for venipuncture. A bag with a steel ball at the junction of the donor tube and bag opening allowed for the ball to be released, which created a vacuum and initiated blood flow. Blood entered a clamped, first bag where it was mixed with anticoagulant. After 500 cc was collected, that blood was centrifuged resulting in a "platelet-rich supernatant plasma" (1200 rpm × 20 min). The clamp was opened, which created an open flow between the first, second (middle), and third bag. The entire assembly was centrifuged again to separate the platelets from the plasma which were packed at the bottom of the third bag (2500 rpm × 40 min). This yielded 60–70% of platelets. Also, with the described three bag technique, packed cells and plasma were isolated simultaneously. This system also allowed for reconstitution of blood with platelet-poor plasma (PPP) to reconstitute platelet depleted blood for various transfusions. Ultimately, this was a closed system that decreased contamination risks of other

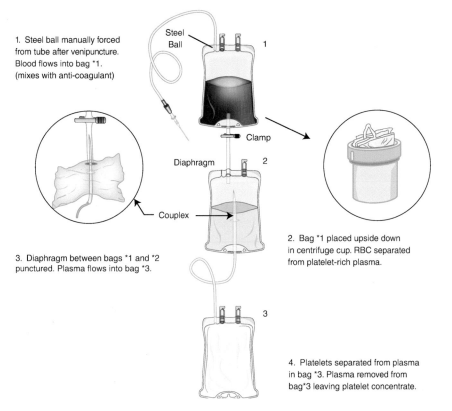

1. Steel ball manually forced from tube after venipuncture. Blood flows into bag *1. (mixes with anti-coagulant)

Steel Ball

1

Clamp

Diaphragm   2

Couplex

3. Diaphragm between bags *1 and *2 punctured. Plasma flows into bag *3.

2. Bag *1 placed upside down in centrifuge cup. RBC separated from platelet-rich plasma.

3

4. Platelets separated from plasma in bag *3. Plasma removed from bag*3 leaving platelet concentrate.

Fig. 1.2 Demonstrates a closed plastic system, which was one of the earlier systems used to isolate platelets

methods. While these advances offered a lot of flexibility for research, clinical use, and blood banks, it was still difficult to obtain large volumes of thrombocytes [3].

Plasmapheresis is one of the greatest developments as a means for large volume platelet isolation. In emergencies that required platelet transfusions, only whole blood had been used for its platelets, which limited its utility due to the amount of volume required. Platelet-rich plasma and platelet concentrates became a clinical focus in this regard. The application of plasmapheresis as a method to generate a large volume of isolated platelets was first introduced in 1961 [22]. The donor's blood flowed into a container which was detached and centrifuged. For this protocol, platelet-rich plasma was obtained by spinning at 1100 rpm for 15 min (Fig. 1.3). Plasma was expressed into a satellite plasma bag, and the red cells were returned directly to the patient with saline, qualifying as plasmapheresis. Following re-transfusion, a second 500 mL whole blood was collected using the second blood container and the process repeated. This yielded 500 mL of platelet-rich plasma which could be collected for transfusion, while packed red blood cells were immediately returned to the donor within the same system (Fig. 1.3). Certainly, the volume of platelets generated by plasmapheresis is

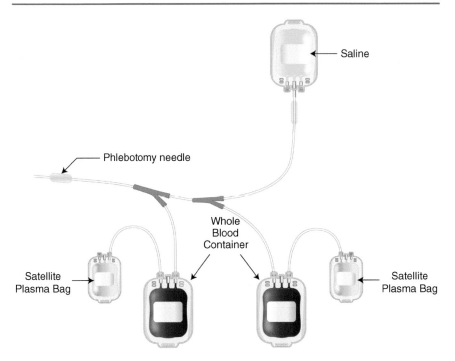

Fig. 1.3 Demonstrates one of the earliest systems to isolate platelets via plasmapheresis

geared towards large-scale patient transfusions, but this was a breakthrough for the patients and donors involved in this treatment. This method was simple, is time effective, and did not involve large investments in devices like the Cohn Blood Fractionator. The volume obtained in the 1961 investigation was two times the Cohn Fractionator's and was capable of producing any amount of plasma from a single individual staged over time, such as 1 L per week for 3 months [22].

While some of the described techniques focused largely on component isolation for transfusions and large volume purposes, it was these scientists' advances that enable us to isolate PRP so conveniently today and to help our patients. Now *400* years after the discovery of blood and circulation, we are still learning about the power and potential of blood, its components, and specifically platelets. The use of platelets and PRT should continue to advance as we better understand and utilize the many components of human blood.

First Uses of PRP

Now that we have a better understanding of how we learned to isolate PRP for present use, we will highlight a few of the high impact original applications in various medical fields.

Bone

In 1994 the idea of utilizing autologous blood products to enhance hard tissue regeneration was first discussed in the oral and maxillofacial surgery (OMS) literature. The investigators isolated fibrinogen from the patient and mixed it with bovine thrombin which was then sprayed over particulate cancellous bone and marrow for reconstruction of mandibular defects [23]. Be more specific here of what the fibrinogen/thrombin spray actually did.

A few years later in 1998, a group of oral and maxillofacial surgeons led by Robert Marx were the first to formally introduce PRP to all fields specifically for the enhancement of bone regeneration [1]. Interestingly, these investigators utilized plasmapheresis in the operating room, as outlined previously in this chapter. Using a cell separator, 400–450 ml of autologous whole blood was obtained through a central venous catheter (CVC). As blood was drawn, citrate phosphate dextrose was added to achieve anticoagulation; whole blood was centrifuged into three components (*RBCs, PRP, PPP*); and ultimately after the PRP was isolated, the remaining components were returned to the patient. What did Marx use the PRP for? Be more specific. This design beautifully demonstrates a novel application for all of the incredible discoveries and innovation dating back to the 1600s. Further, their results suggested that addition of PRP to cancellous cellular marrow accelerated both the rate and degree of bone formation at least for the first 6 months, with a mean trabecular bone area of 74.0% (PRP enhanced) versus 55.1% (control) [1]. This represents a 34% relative increase in bone formation.

I would add a very short paragraph here summarizing the use today of PRP in OMFS.

Skin, Hair, and Wound Healing

PRP for use in the fields of dermatology, cosmetic procedures, and wound healing did not gain widespread popularity in the literature for human use until nearly 2010. Cosmetic applications include facial rejuvenation, facial wrinkles, skin elasticity, and some forms of alopecia [24, 25]. Wound healing benefits have been seen for both treatment of chronic ulcers or scars and accelerated healing postoperatively.

In 2006 an article first described the use of PRP for androgenic alopecia [26]. Investigators delineated two symmetric bald areas on 20 male patients between the ages of 25–55. Follicular units embedded with PRP were implanted compared to untreated follicular units as controls. After a follow-up period of 7 months, the PRP-implanted site exhibited a density of 18.7 follicular units per cm^2, compared to the control group of 16.7 follicular units per cm^2 [26]. This represented a 12% increase in follicular density ($p < 0.001$) [26] and generated tremendous discussion in the field. Since that time, multiple reports appear to support the use of PRP for androgenic alopecia with additional biological and histological support [25].

A median sternotomy is a common surgical access for cardiothoracic (CT) surgery operations. Unfortunately, deep sternal wound infections and dehiscence can affect up to 8% of patients and lead to significant morbidity and even mortality [27]. This clinical problem led the CT surgical field to investigate PRP to promote earlier and improved wound healing. One investigation looked at 2000 patients where 6 mL of PRP was applied to the sternum and soft tissues during closure. This study found an absolute risk reduction of 7.41% for deep and superficial surgical wound infections [27]. Other investigations have also pointed to possible bactericidal properties of PRP [28].

Orthopedics and Arthritis

PRP injections started to be studied for the treatment of osteoarthritis (OA) in the early 2000s. The effect of treatment with intra-articular PRP injections into OA joints were compared to placebo, control, and hyaluronic acid in multiple reports. With regard to both pain and function, the injection of PRP in OA indicates a beneficial therapeutic effect for patients [29, 30]. Although there have been conflicting data, the biology supports its use.

Conclusion

The idea of exploiting an autologous blood product as a therapeutic treatment is now scientifically and clinically accepted. However, it is important to appreciate the challenges and scientific questions that earlier scientists addressed to allow modern-day clinicians to practice the way we do today. It took centuries for blood to be understood, its components to be discovered, device innovations to follow, and then its therapeutic potential explored on animals and humans. Today, in part thanks to the dairy industry, we are able to isolate platelets within 10–20 min from a tabletop centrifuge in the office or operating room.

There is still much to be learned and discovered regarding PRT and its specific factors. The utility of platelets and PRT will only continue to evolve as we move forward in the everlasting quest of scientific advancement. Hopefully, with the available technological resources at our disposal, unrealized questions with respect to PRP will be answered in time for the next chapter.

References

1. Marx RE, Carlson ER, Eichstaedt RM, Schimmele SR, Strauss JE, Georgeff KR. Platelet-rich plasma: growth factor enhancement for bone grafts. Oral Surg Oral Med Oral Pathol Oral Radiol Endod. 1998;85(6):638–46.
2. Keene DJ, Alsousou J, Harrison P, Hulley P, Wagland S, Parsons SR, et al. Platelet rich plasma injection for acute Achilles tendon rupture: PATH-2 randomised, placebo controlled, superiority trial. BMJ. 2019;367:l6132.

3. Klein E, Arnold P, Earl RT, Wake E. A practical method for the aseptic preparation of human platelet concentrates without loss of other blood elements. N Engl J Med. 1956;254(24): 1132–3.
4. Klein E, Farber S, Djerassi I. Conference on obstacles to the control of acute leukemia. Control and prevention of hemorrhage: platelet separation. Cancer Res. 1965;25(9):1504–9.
5. Freeman G. Method for obtaining large yields of human platelets as a by-product of blood collection. Science. 1951;114(2968):527–8.
6. Greenwalt TJ. A short history of transfusion medicine. Transfusion. 1997;37(5):550–63.
7. Silverman ME. William Harvey and the discovery of the circulation of blood. Clin Cardiol. 1985;8(4):244–6.
8. Ellis H. James Blundell, pioneer of blood transfusion. Br J Hosp Med (Lond). 2007;68(8):447.
9. Dzik S. James Blundell, obstetrical hemorrhage, and the origins of transfusion medicine. Transfus Med Rev. 2018;32(4):205–12.
10. Judson G, Jones A, Kellogg R, Buckner D, Eisel R, Perry S, et al. Closed continuous-flow centrifuge. Nature. 1968;217(5131):816–8.
11. Svedberg T, Fåhraeus R. A new method for the determination of the molecular weight of the proteins. J Am Chem Soc. 1926;48(2):430–8.
12. Beams HW, Kessel RG. Development of centrifuges and their use in the study of living cells. In: Bourne GH, Jeon KW, Friedlander M, Jeon KW, editors. International review of cytology, vol. 100. Academic Press; 1987. p. 15–48.
13. Beams JW, Weed AJ, Pickels EG. The ultracentrifuge. Science. 1933;78(2024):338–40.
14. Meselson M, Stahl FW. The replication of DNA in Escherichia coli. Proc Natl Acad Sci U S A. 1958;44(7):671–82.
15. Watson JD, Crick FH. Molecular structure of nucleic acids; a structure for deoxyribose nucleic acid. Nature. 1953;171(4356):737–8.
16. Abel JJ, Rowntree LG, Turner BB. Plasma removal with return of corpuscles (plasmaphaeresis). The Journal of Pharmacology and experimental therapeutics Vol. V. No. 6, July, 1914. Transfus Sci. 1914;11(2):166–77.
17. Freireich EJ, Judson G, Levin RH. Separation and collection of leukocytes. Cancer Res. 1965;25(9):1516–20.
18. Vallee BL, Hughes WL Jr, Gibson JG 2nd. A method for the separation of leukocytes from whole blood by flotation on serum albumin. Blood. 1947;2(1):82–7.
19. Millward BL, Hoeltge GA. The historical development of automated hemapheresis. J Clin Apher. 1982;1(1):25–32.
20. Cohn EJ, editor. Blood collection and preservation. Fourth annual meeting of the American Association of Blood Banks, Minneapolis, Minnesota, October 22–25, 1951.
21. Oon CJ, Hobbs JR. Clinical applications of the continuous flow blood separator machine. Clin Exp Immunol. 1975;20(1):1–16.
22. Kliman A, Gaydos LA, Schroeder LR, Freireich EJ. Repeated plasmapheresis of blood donors as a source of platelets. Blood. 1961;18:303–9.
23. Tayapongsak P, O'Brien DA, Monteiro CB, Arceo-Diaz LY. Autologous fibrin adhesive in mandibular reconstruction with particulate cancellous bone and marrow. J Oral Maxillofac Surg. 1994;52(2):161–5; discussion 6.
24. Zhang M, Park G, Zhou B, Luo D. Applications and efficacy of platelet-rich plasma in dermatology: a clinical review. J Cosmet Dermatol. 2018;17(5):660–5.
25. Leo MS, Kumar AS, Kirit R, Konathan R, Sivamani RK. Systematic review of the use of platelet-rich plasma in aesthetic dermatology. J Cosmet Dermatol. 2015;14(4):315–23.
26. Uebel CO, da Silva JB, Cantarelli D, Martins P. The role of platelet plasma growth factors in male pattern baldness surgery. Plast Reconstr Surg. 2006;118(6):1458–66. discussion 67
27. Patel AN, Selzman CH, Kumpati GS, McKellar SH, Bull DA. Evaluation of autologous platelet rich plasma for cardiac surgery: outcome analysis of 2000 patients. J Cardiothorac Surg. 2016;11(1):62.

28. Bielecki TM, Gazdzik TS, Arendt J, Szczepanski T, Krol W, Wielkoszynski T. Antibacterial effect of autologous platelet gel enriched with growth factors and other active substances: an in vitro study. J Bone Joint Surg Br. 2007;89(3):417–20.
29. Laudy AB, Bakker EW, Rekers M, Moen MH. Efficacy of platelet-rich plasma injections in osteoarthritis of the knee: a systematic review and meta-analysis. Br J Sports Med. 2015;49(10):657–72.
30. Jang SJ, Kim JD, Cha SS. Platelet-rich plasma (PRP) injections as an effective treatment for early osteoarthritis. Eur J Orthop Surg Traumatol. 2013;23(5):573–80.

PRP Vs. PRF

2

Alexander Pascal, Alia Koch, and Regina Landesberg

Introduction

The use of platelet concentrates in the field of regenerative medicine has become more prevalent since its early adoption in dentistry and maxillofacial surgery in 1997 [3, 4]. These concentrates have promising implications for future use with continued research and refinement. Platelets have traditionally been used to treat patients with severe blood loss due to thrombocytopenia. In addition to its involvement with the coagulation cascade, platelets store a significant quantity of growth factors such as platelet-derived growth factor (PDGF), transforming growth factor β-1 (TGF β-1), and vascular endothelial growth factor (VEGF) [5]. Researchers determined that proper isolation techniques could yield high quantities of these growth factors to stimulate the local environment [1, 6]. Platelet-rich plasma (PRP) was first mentioned in publication in 1954, which led to the discovery of its use as a therapeutic agent 10 years later [1].

PRP is a biomaterial that contains high concentrations of platelets and plasma proteins resulting from whole blood centrifugation. The addition of anticoagulants is essential to the production of PRP, which is one of the defining differences between PRP and PRF [7]. While PRP is considerably older than PRF, disparity with the PRP preparation protocol still exists [5] and will be discussed below. Platelets contain numerous secretory granules, with alpha granules containing PDGF, VEGF, TGF-β, insulin-like growth factor (IGF), epithelial growth factor (EGF), endothelial cell growth factor (ECGF), and fibroblast growth factor (FGF). Upon platelet activation and degranulation of alpha granules, the release of growth

A. Pascal · A. Koch (✉)
Division of Oral and Maxillofacial Surgery, NewYork-Presbyterian/Columbia University Irving Medical Center, New York, NY, USA
e-mail: Ak2045@cumc.columbia.edu

R. Landesberg
Private Practice, Oral and Maxillofacial Surgery, Torrington, CT, USA

© The Author(s), under exclusive license to Springer Nature Switzerland AG 2022
E. M. Ferneini et al. (eds.), *Platelet Rich Plasma in Medicine*,
https://doi.org/10.1007/978-3-030-94269-4_2

13

factors and cytokines ultimately accelerates healing via cell proliferation and differentiation [1].

PRF, the second-generation platelet concentrate, was first utilized in 2001 by Chourkroun et al. in the field of oral and maxillofacial surgery [8]. The impetus for the development of PRF emanated from dissatisfaction with the existing PRP preparation protocol and lack of standardization. PRF production is faster, simpler, and more economical compared to its predecessor [9]. Researchers have also sought to improve the regenerative clinical outcomes of platelet concentrates. The advancement from first-generation PRP to second-generation PRF led to modifications in the spatiotemporal delivery system of growth factors. PRF's fibrin matrix is polymerized in a more flexible tetra molecular structure, incorporating higher levels of cytokines and growth factors [7, 10, 11]. The motivation to utilize PRP or PRF remains the same, to leverage the growth factors released for regenerative purposes. While both plasma concentrates contain many of the same growth factors, great interest remains in quantifying the differences in growth factor release and overall efficacy of use [12].

Preparation of PRP and PRF

PRP and PRF preparation protocols utilize the same armamentarium with both requiring centrifugation of whole blood samples to isolate the platelet concentrate. PRP preparation continues to lack any standardized protocol, according to a recent systematic review. PRP techniques currently utilized vary in parameters such as spin time and gravitational force and have led to increased difficulty in comparing the efficacy of treatment across different studies [13, 14]. All PRP protocols begin with blood sample collection with anticoagulant and immediate centrifugation for a set time point. The result is three separate layers with the red blood cells located on the bottom, the acellular plasma located on top, and the "buffy coat" located in the middle. Depending on the PRP protocol utilized, a second round of centrifugation is utilized to isolate the "buffy layer," which contains the concentrated platelets. Once the layer is isolated, thrombin and calcium chloride are utilized to activate the platelets and enable fibrin polymerization as the concentrates are applied to the site of targeted therapy [5].

PRF fabrication can be achieved with a more simplistic approach in comparison to PRP. It should be noted that PRF does not require the addition of anticoagulants, thrombin, or calcium chloride following sample collection. Physiologic levels of thrombin are sufficient to enable the formation of a fibrin matrix [7]. The blood sample is collected and immediately centrifuged resulting in three layers: the top layer consisting of cellular plasma, the middle layer consisting of the PRF clot, and the bottom layer consisting of red blood cells. Contrary to PRP, the absence of anticoagulants in the PRF protocol results in platelet activation and the coagulation process upon contact with the glass surface of the sample tube. Fibrinogen is primarily located in the upper portion of the tube, ultimately coming into contact with thrombin, resulting in a fibrin product [6].

The preparation protocol for PRF is cited as one of the main advantages over the use of PRP. PRF fabrication is simplistic and low cost in nature, making it a more appealing product when compared to PRP. PRF does not require any manipulation of the blood sample unlike PRP, which requires anticoagulant and thrombin additives. There is a reduced chance for user error when fabricating PRF as there are fewer steps involved and no blood manipulation [6, 7]. Much controversy and variation exists when discussing the PRP preparation protocol, creating difficulty when comparing the efficacy of PRP across studies [13]. It is worth mentioning that the PRF preparation is time sensitive; therefore the user must be well versed in the protocol and blood collection. In addition, PRP does not require the use of glass coated tubes, which is needed for clot polymerization in the PRF protocol [6, 10, 15].

Efficacy of Growth Factor Release and Use in a Clinical Setting

PRF, a second-generation platelet concentrate, is lauded for its simplistic approach when compared to the first-generation platelet concentrate PRP. This simplistic approach is a driving force behind PRF's utilization over PRP, but researchers continue to investigate the efficacy of growth factor release and overall clinical results [16]. In a study involving 18 blood samples from 6 donors, the authors aimed to compare the growth factor release from PRP, PRF, and a modernized PRF product identified as advanced platelet-rich fibrin (A-PRF) [12]. A-PRF utilizes a modified preparation protocol in comparison to traditional PRF by reducing the centrifugation speed from 2700 rpm to 1500 rpm. Curtailing the centrifugation speed has previously shown a resulting increase in platelet quantity and improvement in autologous cell function [17]. The release of growth factors PDGF-AA, PDGF-AB, PDGF-BB, TGFB1, VEGF, EGF, and IGF was quantified using ELISA at time points of 15 min, 60 min, 8 h, 1 day, 3 days, and 10 days. At earlier time points, PRP was found to release a greater number of growth factors when compared to PRF and A-PRF. However, PRF and A-PRF were found to release a larger number of growth factors released at later time points over PRP. A-PRF exhibited the most impressive results not only regarding growth factor release at the later time points but also with total protein accumulation over the length of the study [12].

The prolonged release of growth factors and cytokines by PRF does not necessarily translate to superior results over PRP in a clinical setting. A meta-analysis of randomized clinical trials comparing the efficacy of PRF and PRP in arthroscopic rotator cuff repair found improved clinical results with PRP utilization over PRF. The authors found that PRP had clear benefits in accelerating tendon healing rates, pain level, and overall functional outcome. No benefit to tendon healing or functional outcomes was seen with PRF utilization [18]. A prospective clinical study on PRF use in rotator cuff tendon healing went so far as to suggest PRF inhibits the tendon healing process. It was hypothesized that following the dissolution of the fibrin clot sutured between tendon and bone, a space forms which is detrimental to therapeutic efforts [19].

One can surmise that the clinical scenario is a key determinant in choosing between platelet concentrates. As previously mentioned, a meta-analysis of randomized clinical trials found improved clinical outcomes with PRP utilization for arthroscopic rotator cuff repair [18]. However, studies also demonstrated that PRF's controlled long-term release of growth factors improved osteoblastic proliferation and differentiation when compared (to) PRP [11]. A comparative study of PRP and PRF on rat osteoblasts explains PRP's overall limited potential in stimulating bone regeneration. PRP's rapid growth factor release prior to local tissue cell outgrowth and thrombin's potential toxic effects are thought to reduce PRP's regenerative potential. Contrarily, the gradual release of growth factors with PRF utilization was said to lead to a stronger and more sustained osteoblastic response in rats in vitro [11]. The treatment of soft tissue defects with PRF has been shown to have improved healing and regenerative potential. PRF is found to increase the amount of keratinized mucosa surrounding dental implants, and maintaining a specified width of keratinized mucosa is thought to be a determinant in periodontal health and the prevention of gingival recession [20].

Conclusion

There is mounting evidence regarding the therapeutic potential of PRP and PRF. However, discrepancies continue to exist in scientific literature when comparing the efficacy of treatment in a clinical setting. The lack of standardization in the PRP preparation protocol is thought to be a contributing factor to these inconsistencies. PRF's simplistic preparation protocol and gradual sustained release of growth factors are major advantages over PRP. However, the supposition that PRF will provide superior results as a therapeutic agent does not consistently translate into clinical trials. Additional, high-quality studies are needed to better interpret the efficacy of use across the various applications in regenerative medicine.

References

1. Scully D, Naseem KM, Matsakas A. Platelet biology in regenerative medicine of skeletal muscle. Acta Physiol. 2018;223(3):e13071.
2. Kim T-H, Kim S-H, Sándor GK, Kim Y-D. Comparison of platelet-rich plasma (PRP), platelet-rich fibrin (PRF), and concentrated growth factor (CGF) in rabbit-skull defect healing. Arch Oral Biol. 2014;59(5):550–8.
3. Hall MP, Band PA, Meislin RJ, Jazrawi LM, Cardone DA. Platelet-rich plasma: current concepts and application in sports medicine. J Am Acad Orthop Surg. 2009;17(10):602–8.
4. Marx RE, Carlson ER, Eichstaedt RM, Schimmele SR, Strauss JE, Georgeff KR. Platelet-rich plasma: growth factor enhancement for bone grafts. Oral Surg Oral Med Oral Pathol Oral Radiol Endod. 1998;85(6):638–46.
5. Dohan Ehrenfest DM, Rasmusson L, Albrektsson T. Classification of platelet concentrates: from pure platelet-rich plasma (P-PRP) to leucocyte- and platelet-rich fibrin (L-PRF). Trends Biotechnol. 2009;27(3):158–67.

6. Borie E, Olíví DG, Orsi IA, Garlet K, Weber B, Beltrán V, Fuentes R. Platelet-rich fibrin application in dentistry: a literature review. Int J Clin Exp Med. 2015;8(5):7922–9.
7. Giannini S, Cielo A, Bonanome L, Rastelli C, Derla C, Corpaci F, Falisi G. Comparison between PRP, PRGF and PRF: lights and shadows in three similar but different protocols. Eur Rev Med Pharmacol Sci. 2015;19(6):927–30.
8. Choukroun J, Adda F, Schoeffler C, Vervelle A. Une opportunité en paro-implantologie: le PRF. Implantodontie. 2000;42:55–62.
9. Caruana A, Savina D, Macedo JP, Soares SC. From platelet-rich plasma to advanced platelet-rich fibrin: biological achievements and clinical advances in modern surgery. Eur J Dent. 2019;13(02):280–6.
10. Saluja H, Dehane V, Mahindra U. Platelet-rich fibrin: a second generation platelet concentrate and a new friend of oral and maxillofacial surgeons. Ann Maxillofac Surg. 2011;1(1):53–7.
11. He L, Lin Y, Hu X, Zhang Y, Wu H. A comparative study of platelet-rich fibrin (PRF) and platelet-rich plasma (PRP) on the effect of proliferation and differentiation of rat osteoblasts in vitro. Oral Surg Oral Med Oral Pathol Oral Radiol Endod. 2009;108(5):707–13.
12. Kobayashi E, Flückiger L, Fujioka-Kobayashi M, Sawada K, Sculean A, Schaller B, Miron RJ. Comparative release of growth factors from PRP, PRF, and advanced-PRF. Clin Oral Investig. 2016;20(9):2353–60.
13. Kramer ME, Keaney TC. Systematic review of platelet-rich plasma (PRP) preparation and composition for the treatment of androgenetic alopecia. J Cosmet Dermatol. 2018;17(5):666–71.
14. Chahla J, Cinque ME, Piuzzi NS, Mannava S, Geeslin AG, Murray IR, Dornan GJ, Muschler GF, LaPrade RF. A call for standardization in platelet-rich plasma preparation protocols and composition reporting: a systematic review of the clinical orthopaedic literature. J Bone Joint Surg Am. 2017;99(20):1769–79.
15. Masuki H, Okudera T, Watanebe T, Suzuki M, Nishiyama K, Okudera H, Nakata K, Uematsu K, Su C-Y, Kawase T. Growth factor and pro-inflammatory cytokine contents in platelet-rich plasma (PRP), plasma rich in growth factors (PRGF), advanced platelet-rich fibrin (A-PRF), and concentrated growth factors (CGF). Int J Implant Dent. 2016;2(1):19.
16. Li Q, Pan S, Dangaria SJ, Gopinathan G, Kolokythas A, Chu S, Geng Y, Zhou Y, Luan X. Platelet-rich fibrin promotes periodontal regeneration and enhances alveolar bone augmentation. Biomed Res Int. 2013;2013:638043.
17. Ghanaati S, Booms P, Orlowska A, Kubesch A, Lorenz J, Rutkowski J, Landes C, Sader R, Kirkpatrick C, Choukroun J. Advanced platelet-rich fibrin: a new concept for cell-based tissue engineering by means of inflammatory cells. J Oral Implantol. 2014;40(6):679–89.
18. Hurley ET, Lim Fat D, Moran CJ, Mullett H. The efficacy of platelet-rich plasma and platelet-rich fibrin in arthroscopic rotator cuff repair: a meta-analysis of randomized controlled trials. Am J Sports Med. 2019;47(3):753–61.
19. Rodeo SA, Delos D, Williams RJ, Adler RS, Pearle A, Warren RF. The effect of platelet-rich fibrin matrix on rotator cuff tendon healing: a prospective, randomized clinical study. Am J Sports Med. 2012;40(6):1234–41.
20. Temmerman A, Cleeren GJ, Castro AB, Teughels W, Quirynen M. L-PRF for increasing the width of keratinized mucosa around implants: a split-mouth, randomized, controlled pilot clinical trial. J Periodontal Res. 2018;53(5):793–800.

PRP Preparation

3

Steven Halepas and Regina Landesberg

Introduction

The blood contains numerous cells, proteins, and factors that aid in wound repair. Platelets are cytoplasmic fragments of megakaryocytes and are the main players in primary wound healing. Platelets are known to secrete many protein growth factors that aid in wound healing. Platelets secrete high quantities of growth factors and cytokines such as platelet-derived growth factor (PDGF), transforming growth factor beta-1 (TGFβ1), and vascular endothelial growth factor (VEGF), among others. Primary wound healing is essential in clinical medicine, and without it, surgical procedures would not be possible. Primary hemostasis starts with the activation/aggregation stage. An injury or surgical procedure damages endothelium resulting in exposure of collagen and tissue factor. Integrin GPIa-IIa and GPVI bind to collagen first, and von Willebrand factor (vWF) binds to the expose collagen. The circulating platelets bind to the vWF via GP1b-IX-V receptor. The platelets change shape producing many filopodia to increase their surface contact, and their receptors become active by P2Y1 and ADP. Activated platelet turns on scramblase, moves negatively charged phospholipids from the inner to the outer platelet membrane surface, and provides catalytic surface for tenase (FXa) and prothrombinase complex (FXa + FVa) using calcium ions as glue. The activation increases TxA_2 production which stimulate platelet activation on other platelets as well as its own thromboxane receptors. The platelets then secrete their granules. Dense granules include ADP, which, we stated, activate platelets and help them bind to endothelium

S. Halepas (✉)
Division of Oral and Maxillofacial Surgery, NewYork-Presbyterian/Columbia University Irving Medical Center, New York, NY, USA
e-mail: sh3808@cumc.columbia.edu

R. Landesberg
Private Practice, Oral and Maxillofacial Surgery, Torrington, CT, USA

© The Author(s), under exclusive license to Springer Nature Switzerland AG 2022   19
E. M. Ferneini et al. (eds.), *Platelet Rich Plasma in Medicine*,
https://doi.org/10.1007/978-3-030-94269-4_3

and cause glycoprotein 2b and 3a (GpIIb/IIIa) to be expressed on platelets surface. Alpha granules contain PDGF, TGFβ, fibrinogens, vWFs, and others. Fibrinogen binds to glycoprotein 2b and 3a receptor allowing linkage of platelets to occur. This lays the foundation for secondary hemostasis to occur.

Clearly, these tiny cytoplasmic fragments are intensely complicated, and their natural role in wound healing makes them extremely potential to aid in post-surgical and clinical applications. Marx first described the use of platelet-rich plasma (PRP) and platelet-rich fibrin (PRF) in the dental field in 1998 where he reported positive healing of the alveolar bone with its use [1]. In 2001, Marx attempted to define PRP. He stated: "normal platelet counts in blood range between 150,000/μL and 350,000/μL and average about 200,000/μL. Because the scientific proof of bone and soft tissue healing enhancement has been shown using PRP with 1,000,000 platelets/μL, it is this concentration of platelets in a 5-mL volume of plasma which is the working definition of PRP today" [2]. Differentiation in preparation results in different platelet concentrations. The easiest way to remember it is at least one million platelets per microliter in 5-mL product.

All PRP/PRF preparation techniques begin with phlebotomy and collecting blood. The blood is placed in a centrifuge to separate into the red blood cells layer at the bottom, the acellular plasma layer at the top, and the middle platelet concentrate layer in the middle. The protocol for preparation varies depending on which type of platelet-rich product the practitioner is hoping to achieve. Ehrenfest et al. attempted to define four main families of preparations the are delineated based on their cellular content [3, 4] (see Fig. 3.1). They define the pure platelet-rich plasma (P-PRP) as leucocyte poor platelet-rich plasma product that is made without leucocytes and with a low-density fibrin network after activation. The second category is leucocyte- and platelet-rich plasma (L-PRP) which is the most utilized in the literature and clinical practice as most commercial kits fabricate this type. It is similar to the first with the addition of leucocytes. The third category is pure platelet-rich fibrin (P-PRF) which is without leucocytes but has a high-density fibrin network that produced a strong gel clot. The last category they describe is leucocyte- and platelet-rich fibrin (L-PRF).

Commercial Products

The protocol for platelet-rich product production varies depending on which classification is needed. Many commercial products are available that include specialized kits and centrifuges that are calibrated to generate the desired product and with high concentrations. The reason the commercial products have become so popular is due to their usability, requiring almost no outside knowledge by the user, and the consistency of the platelet concentrations. The drawback to these products is the cost. While a simple centrifuge can be used to obtain similar results, without active knowledge by the clinical, the final product might not have platelet concentrations

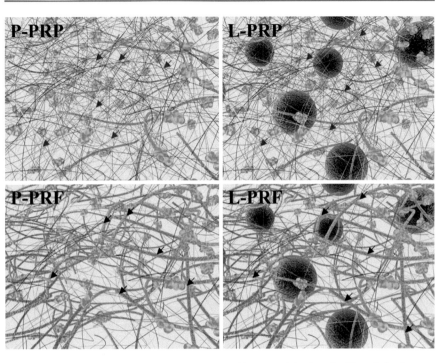

Fig. 3.1 This is an illustration of the matrix and architecture of the four families of platelet concentrations described by Ehrenfest et al. (Reproduced with permission and without alteration from Dohan Ehrenfest et al. [3], Figure 4)

above baseline. Remember, as defined by Marx, platelet-rich plasma is a volume of autologous plasma that has a platelet concentration above baseline [2]. One study compared six different single-spin methods of preparation that are common in the private practice. They found that all six produced PRP with moderate variability in concentration and two methods were able to produce P-PRP (gel tube separator and double-syringe Arthrex ACP®). The cost for supplies varies dramatically, with the lowest being about $2 and the most expensive around $234 [5].

After doing a quick literature search, the most common commercial brands for the fabrication of platelet-rich products included Harvest Technologies, Arteriocyte Magellan, Cytomedix Angel System (Arthrex), GenesisCS APC, RegenKit, Biomet GPS III Platelet Separation System, PCCS from Biomet 3i, Squire, CATS, Sequestra, and UltraConcentrator. It is very likely that many other companies exist that make equivalent products. The commercial systems are easy to use, and the practitioner should follow the manufacturer's specific protocol. This chapter will focus on the protocol for PRF and PRP fabrication using any basic centrifuge and not a commercial kit.

Protocol for PRF Preparation

PRF is more simplistic then that of PRP production. To fabricate PRF, the provider must obtain 10 cc of the patient's blood in an appropriate blood sample tube, typically a red top single-use BD Vacutainer® blood tubes with no added preservatives. The sample should then be immediately placed in a centrifuge and run at 2,500 rpm for 12 min. Failure to immediately place the tube in the centrifuge can inhibit PRF formation. One study recommends the fabrication of what they call advanced PRF. Production is made similarly to the above mentioned PRF but instead centrifuged at a slower speed of 1500 rpm for 14 min [6]. The modification is believed to increase the platelet cell numbers and macrophage behavior. The yellow clot layer can then be removed from the tube using college pliers (see Fig. 3.2).

Fig. 3.2 PRF after being removed from a 10 cc BD Vacutainer®

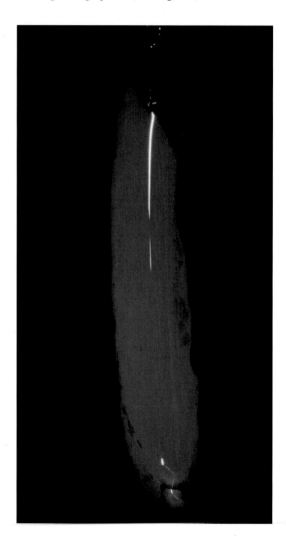

Protocol for PRP Preparation

The protocol for PRP preparation has much more variability than the production of PRF (see Table 3.1). PRF can be easily made from a standard centrifuge. Many commercially available PRP kits exist with specific centrifuges to make the process easier, although this can contribute cost. An alternative to this is a process described by Ben Eby. First 8 cc of acid-citrate-dextrose solution A (ACD-A) is drawn into a 60 cc syringe. Next 52 cc of venous blood is drawn though the IV site into the syringe. Gently rock the vials back and forth to allow mixing of the ACD-A and the blood. Draw 9 cc of the ACD-A/blood mixture and place into a 10 cc BD Vacutainer® and repeat for a total of six tubes. (If drawing into 10 cc BD Vacutainer®, draw 8 cc of venous blood and add 1.2 cc of ACD-A.) The tubes are placed in to the centrifuge and run at 3000 rpms for 10 min. Next, using a 30 mL syringe and a 16-gauge catheter, draw off the yellow top layer from each tube leaving the compact red blood cell layer behind. When ready to use the PRP, one must activate it, using 5000 units of bovine thrombin powder and 5 cc of 10% CaCl. Next mix 0.15 mL of the thrombin/CaCl to each millimeter of PRP you wish to activate. While the utilization of more expensive centrifuges (upwards of $8000 and $250 single-use components) may yield larger platelet concentrations, Dr. Eby reports a 358% increase with the described method [7].

An alternative method is referred to as the ACD yellow top tube method after Petersons and Reeves. With this protocol, 8.5 cc of whole blood is collected in an ACD yellow top tube (BD Vacutainer ACD®). The tube is placed in a centrifuge and spun at 1000 g [RCF (g) = 1.12 × radius × (rpm/1000)^2] for 10 min. The platelet-rich poor layers is drown off from the top leaving 1–2 cc of PRP just below [8].

Table 3.1 The cost, time, and PRP produced from different PRP protocols

Method	Supply cost (estimate)	Time (min)	Whole blood used (mL)	PRP volume produced (mL)	Platelet concentration per 5 mL of PRP	Total platelet concentration per protocol
ACD yellow top tube	$2	15	8	2.1	3,047,615	1,279,998
Single syringe	$3	20	14	4.1	3,089,020	2,532,996
Eby method	$55	15	52	7	4,370,900	6,119,260
YCellBio blood separation kit	$62	15	15	4	3,592,500	2,874,000
Eclipse PRP kit	$102	15	11	12.7	3,579,625	9,092,247
Arthrex ACP double-syringe system	$113	10	15	5.6	2,817,855	1,279,998
Arthrex angel	$234	40	120	17	5,213,235	2,532,996

Data to construct this table was obtained from Harrison et al. [5]

A third protocol is referred to as the single 20 cc syringe method by Harrison et al. The authors state 1.5 cc of sodium citrate at a concentration of 40 mg/mL is drawn up into a standard plastic 20 cc syringe. 15 cc of whole blood is then drawn into the syringe and mixed. The syringe is placed into a centrifuge for 10 min at 1000 g [5]. All the plasma layer and then 0.6 cc of the red layer is included. For comparison, Harrison et al. have a table that includes the single syringe and ACD yellow top tube with three commercially available products: Eclipse PRP Kit, YCellBio Blood Separation Kit, Arthrex Angel, and Arthrex ACP Double-Syringe System.

Conclusion

Most classify platelet-rich products into four main families of plasma products: P-PRP, L-PRP, P-PRF, and L-PRF. Different commercial products that include a commercial branded centrifuge with single use components are available that can produce each of the four families. Each manufacture includes a specific protocol in terms of how much blood is needed. While the optimal PRP dosage for most clinical applications is unknown, the available literature does suggest an inadequate and excessive dosage exists, suggesting an optimal dose can be found with further investigation. PRP and PRF can be fabricated with a simple single-spin centrifuge and without expensive commercial cartridges or components, but the levels of bioactive product are likely to be more inconsistent. Providers should perform a cost/benefit analysis to determine if one of the commercially available systems would be applicable in their daily practice and for their specific patient care needs.

References

1. Marx RE, Carlson ER, Eichstaedt RM, Schimmele SR, Strauss JE, Georgeff KR. Platelet-rich plasma: growth factor enhancement for bone grafts. Oral Surg Oral Med Oral Pathol Oral Radiol Endod. 1998;85(6):638–46.
2. Marx RE. Platelet-rich plasma (PRP): what is PRP and what is not PRP? Implant Dent. 2001;10(4):225–8.
3. Dohan Ehrenfest DM, Rasmusson L, Albrektsson T. Classification of platelet concentrates: from pure platelet-rich plasma (P-PRP) to leucocyte- and platelet-rich fibrin (L-PRF). Trends Biotechnol. 2009;27(3):158–67. https://doi.org/10.1016/j.tibtech.2008.11.009. Epub 2009 Jan 31. PMID: 19187989.
4. Dohan Ehrenfest DM, Bielecki T, Mishra A, Borzini P, Inchingolo F, Sammartino G, et al. In search of a consensus terminology in the field of platelet concentrates for surgical use: platelet-rich plasma (PRP), platelet-rich fibrin (PRF), fibrin gel polymerization and leukocytes. Curr Pharm Biotechnol. 2012;13(7):1131–7.
5. Harrison TE, Bowler J, Levins TN, Cheng AL, Reeves KD. Platelet yield and yield consistency for six single-spin methods of platelet rich plasma preparation. Platelets. 2020;31(5):661–6.
6. Kobayashi E, Fluckiger L, Fujioka-Kobayashi M, Sawada K, Sculean A, Schaller B, et al. Comparative release of growth factors from PRP, PRF, and advanced-PRF. Clin Oral Investig. 2016;20(9):2353–60.
7. Eby BW. Platelet-rich plasma: harvesting with a single-spin centrifuge. J Oral Implantol. 2002;28(6):297–301.
8. Peterson N, Reeves KD. Efficacy of one day training in low-cost manual preparation of high cellular platelet rich plasma. J Prolother. 2014;6:e922–e7.

PRP in Orthopedics

4

Nicholas Genovese, Michael Metrione, Lea Bach,
Katherine Lauritsen, and Sheldon S. Lin

Introduction to PRP in Orthopedics

PRP is an autologous growth factor therapy containing a platelet concentration greater than that of whole blood. Centrifugation and separation techniques are used to prepare PRP [1]. PRP accelerates tissue regeneration via three distinct mechanisms: anabolic, anti-inflammatory, and scaffolding. In the anabolic mechanism, platelets secrete growth factors including TGF-β, IGF-1, and VEGF. These growth factors promote mitosis, angiogenesis, and cell migration leading to increased tissue regeneration. In the anti-inflammatory mechanism, platelets release anti-inflammatory cytokines such as IL-1 receptor antagonist, IL-4, IL-10, and IL-13 which preserve cartilage indirectly by limiting the catabolic effects of pro-inflammatory cytokines [2, 3]. The scaffolding mechanism is triggered by PRP's plasma proteins. Fibrinogen along with other clotting factors form a fibrin matrix that allows for cell adherence and proliferation [3].

The composition of PRP varies due to its autologous nature and differences in preparation methods [4, 5]. While they primarily contain platelets, PRP samples may now include leukocytes and thrombin or calcium chloride for platelet activation [1, 5]. Numerous PRP preparations can be used clinically because the FDA exempts autologously derived blood products from regulation. Furthermore, many preparation kits with different platelet recovery capacities have entered the market via the FDA's 510(k) application [6]. The resulting disparity among PRP samples reduces the credibility of generalizations about PRP from literature.

N. Genovese · M. Metrione · L. Bach · K. Lauritsen · S. S. Lin (✉)
Department of Orthopaedics, Rutgers New Jersey Medical School, Newark, NJ, USA
e-mail: linss@njms.rutgers.edu

Osseous Healing

PRP is commonly mixed with bone grafts to enhance the grafts' effects, as seen in Gu et al. [7]. In this study, a PRP scaffold was administered in conjunction with a cancellous bone autograft to 14 patients with Hepple stage V osteochondral lesions (OCLs) of the talus. Multiple patient follow-ups were conducted for up to 2 years post-surgery. During these follow-ups, ankle joint range of motion (ROM) and general health, assessed by the VAS and SF-36, were regularly recorded. Ankle radiographs and an MRI were gathered at the final follow-up. The MRI showed a full regeneration of subchondral bone and cartilage in all patients. Radiographs displayed a bony union at the osteotomy site in all patients. Furthermore, ROM and general health improved significantly when comparing pre- and postoperative scores ($p < 0.0001$ for both). Healing occurred without complications in all patients.

Görmeli et al. [8] compared the efficacy of PRP and hyaluronic acid (HA) injections as adjuncts to arthroscopic microfracture surgery on OCLs of the talus. 40 patients were selected for the study and randomly divided into 3 groups: a PRP group ($n = 13$), an HA group ($n = 14$), and a saline control group ($n = 13$). All patients were initially treated surgically using arthroscopic debridement and microfracture techniques. PRP, HA, or saline injections were administered 24–36 h after surgery. Pain and function were evaluated prior to treatment and at a follow-up (mean = 15.3 months, range = 11–25 months) using VAS and American Orthopaedic Foot and Ankle Society (AOFAS) scoring. A decrease in VAS score demonstrates improvement in health and well-being, whereas an increase in AOFAS score signifies lesser pain and greater function. Postoperative VAS scores were significantly lower in all three groups compared to preoperative values ($p < 0.001$ for all groups). In particular, the PRP group showed a significantly greater decrease in VAS value than the HA or control group's ($p < 0.005$). For AOFAS scoring, follow-up scores were significantly higher than baseline values for all three groups ($p < 0.001$ for all groups). The increase in AOFAS score for the PRP group was significantly greater than the increase in HA or control group value ($p < 0.005$).

Wei et al. [9] determined whether an allograft combined with PRP would improve displaced intra-articular calcaneal fracture healing. 276 fractures were randomly selected to receive one of three treatments: an autograft alone ($n = 101$), an allograft alone ($n = 90$), or an allograft with PRP ($n = 85$). Follow-ups were conducted at 12, 24, and 72 months post-surgery. Healing outcomes were evaluated by AOFAS scoring and radiographic parameters (Bohler's angle, Gissane's angle, calcaneal body dimensions). At 24 and 72 months, the allograft with PRP and autograft groups had significantly better AOFAS scores and radiographic parameters than those of the allograft alone group ($p < 0.05$). No significant differences in outcomes were found between the allograft with PRP and the autograft (see Table 4.1).

Table 4.1 PRP for osseous healing

Study (level of evidence)	Procedure and disease	PRP preparation method (kit manufacturer)	Application	Activation method	Combined treatments	Comparison group	Outcome
Gormeli et al. [8] (I)	OCLs of the talus	(SmartPrep 2 system, harvest autologous hemobiologics)	Single injection occurred 24–36 h after surgery	No	Arthroscopic debridement and microfracture techniques	Yes (HA and saline)	Pain and function in PRP group were significantly better than in HA or saline groups
Wei et al. [9] (I)	Displaced intra-articular calcaneal fractures	Double centrifugation (laboratory made)	Surgical	10% calcium chloride and bovine thrombin solution	Allograft	Yes (ABG, allograft alone)	Function and radiographic parameters in allograft with PRP group were significantly greater than in allograft alone group

Ankle Osteoarthritis

Ankle osteoarthritis is a relatively infrequent degenerative joint disorder with limited treatment options [10]. Surgery is mainly preferred for late-stage and advanced cases. Less invasive options, including PRP injections, are being researched to treat mild to moderate forms of ankle osteoarthritis [11].

Fukawa et al. [11] examined the effectiveness of biweekly PRP injections in 20 patients with ankle osteoarthritis. One ankle from each patient was treated with three injections of 2 mL PRP during each session. Injections were guided using ultrasound. Patients were evaluated at 4, 12, and 24-week time points using VAS and a Self-Administered Foot Evaluation Questionnaire (SAFE-Q). An increase in SAFE-Q score indicates improvement in overall health. VAS scores at each follow-up time point were significantly lower than baseline (baseline vs 4 weeks: $p = 0.06$, baseline vs 12 weeks: $p < 0.001$, baseline vs 24 weeks: $p = 0.02$). SAFE-Q scores were higher than baseline with significance only at the 12-week time point (baseline vs 12 weeks: $p = 0.04$). Though the results support PRP's use in modulating osteoarthritic pain, VAS scores and SAFE-Q scores peaked at 12 weeks signifying that PRP's effects were short term in this study.

A case series by Repetto et al. [10] determined whether PRP injections given to 20 patients with medium to advanced ankle osteoarthritis would improve symptoms and delay surgery. Previously, patients had tried other therapies for at least 6 months with no improvement. In this study, one ankle from each patient was treated with four injections of 3 mL PRP once a week. Follow-ups were conducted at a mean of 17.7 ± 6.4 months post-therapy. Pain and function were assessed using VAS and Foot and Ankle Disability Index (FADI), respectively. A significant reduction in pain ($p = 0.0001$) and a significant improvement in ankle function ($p = 0.001$) were found from baseline to follow-up for 18 of 20 patients. Two patients had to drop out early and needed surgery because the treatment was inadequate for them. Although this level IV case series concluded that weekly PRP injections can improve mid- to long-term symptoms of patients with late-stage ankle osteoarthritis, this study is limited due to small sample size and lack of a control.

Introduction to PRP in Sports Medicine

Platelet-rich plasma (PRP) is an autologous substance rich in platelets that release a wide variety of growth factors from both α and dense granules. PRP contains PDGF, vascular endothelial growth factor, TGF-β, IGF-1, b-FGF, human growth factor, and endothelial growth factor. These growth factors have been associated with the initiation of a healing cascade resulting in cellular chemotaxis, angiogenesis, collagen matrix synthesis, and cell proliferation [5, 11]. Interest in the use of PRP for the treatment and adjunct to sports medicine pathology is increasing; however, its use remains controversial in some regard. Media portrayal of positive results of PRP to treat professional athlete injuries has garnered attention from elite and recreation athletes alike [12]. The use of PRP by athletes is legal in all US professional sports leagues, as well as the World Anti-Doping Agency, and has gained popularity by athletes to treat conditions such as rotator cuff tears, tendinopathies, and ligamentous injury [13]. As previously discussed, platelets house several growth factors including transforming growth factor beta (TGF-B), platelet-derived growth factor (PDGF), insulin-like growth factors (IGF I, II) fibroblast growth factor (FGF), epidermal growth factor (EGF), and vascular endothelial growth factor (VEGF) [14]. These growth factors have been associated with the initiation of a healing cascade resulting in cellular chemotaxis, angiogenesis, collagen matrix synthesis, and cell proliferation [5, 11].

The challenge that sports medicine clinicians face is interpreting the results of the many studies examining the effects of PRP injection due to the significant heterogeneity in PRP preparations and injuries that are attempting to be treated. The following sections will highlight the use of PRP in common specific pathologies encountered by the sports orthopedic clinician.

Muscle Injuries

Muscular injuries account for a large proportion of sports-related injuries, often leading to prolonged absence from sport and lengthy rehabilitation. Of these lesions, most occur in the major muscle groups: hamstrings, adductors, quadriceps, and calf muscles, with hamstring strain accounting for about 29% of all sports injuries and recurrence rates up to 40% within the first year [15–17]. Various treatment algorithms exist focused on returning patients to a pre-injury level of play, in as little time as possible with minimal risk of recurrence [18]. Despite the prevalence and recurrence of these injuries, little evidence exists to support specific management protocols [19, 20].

The use of PRP for muscle injuries has grown considerably over the last decade, based on the effects that growth factors have on stimulating tissue healing and accelerating myofiber regeneration [21, 22]. While early in vitro and animal studies evaluating the efficacy of PRP on muscle healing showed a beneficial effect, these results have not consistently resulted in improvements in humans, bringing into quest the actual clinical benefit [23–25].

Hammond et al. [25] evaluated the effectiveness of PRP in a rat injury model with induced tibialis anterior muscle strains. They found that rats who received PRP injections at site of muscle strain had a quicker recovery as well as a quicker return of pre-injury contractile function. In 2014, Hamid et al. [26] performed a randomized controlled trial assessing short-term results of PRP injections plus rehabilitation protocol compared with rehabilitation alone in 28 athletes with acute hamstring injuries. Subjects in the PRP group had faster return to play than subjects in the rehabilitation-only group (26.7 ± 7.0 versus 42.5 ± 20.6 days) and less pain. Similarly, Rossi et al. [27] performed a randomized controlled trial evaluating PRP injections plus rehabilitation compared with rehabilitation alone with 2-year follow-up. Patients in the PRP group achieved quicker full recover (21.1 ± 3.1 days versus 25 ± 2.8 days) with significantly lower pain severity scores. Difference in recurrence rates after 2-year follow-up was not statistically significant between groups.

Table 4.2 PRP for muscle injuries

Study	Year	Study type	Control group	Double/ single blind	US guided	Return to play mean (days) PRP	Return to play mean (days) control	Reinjury PRP vs control ratio
Reurink et al. [29]	2015	RCT	Saline injection	Double blind	Yes	43.6 +/− 20.2	47.9 +/− 23.3	0.9:1
Hamilton et al. [30]	2015	RCT	None or PPP injection	Single blind	Yes	22.2 +/− 9.6	25.2 +/− 8.8	0.75:1
Rossi et al. [27]	2016	RCT	None	Single blind	Yes	21.1 +/− 3.5	25 +/− 2.8	0.5:1
Martinez-Zapata et al. [31]	2016	RCT	Hematoma evacuation	Double blind	Yes	31.6 +/− 15.4	38.4 +/− 18.6	1:1

*Data using PPP as control was excluded
PPP platelet poor plasma

Despite early promising outcomes, the benefits were less consistent as higher-level studies were completed (see Table 4.2). In 2015, Pas et al. [28] performed a meta-analysis evaluating all literature investigating the use of PRP in acute hamstring injuries. The interpretation of these results was that there was no significant difference when evaluating for pain, return to play, recurrence rates, or strength when compared with study controls. Based on the literature, potential for PRP exists to play a beneficial role in treating muscular injuries, but at this time, there remains insufficient evidence to support its use in standard treatment protocols.

Platelet-Rich Plasma for Achilles Tendinopathy

Achilles tendinopathy is a common musculoskeletal complaint and often occurs due to overuse of the Achilles tendon [32, 33]. PRP is one of several conservative treatments applicable to this injury. Research on PRP's value as a treatment for Achilles tendinopathy is growing because no single treatment has emerged as the best therapy for this condition [33].

Zou et al. [32] investigated PRP's efficacy on ruptured Achilles tendon healing. The study consisted of 36 patients randomly divided into two groups: a PRP group ($n = 16$) in which PRP was injected into the paratenon sheath during surgery and a control group ($n = 20$) in which no injection took place. Postoperatively, patients had multiple follow-ups for up to 2 years during which the Leppilahti score, SF-36 score, ankle ROM, and calf strength were assessed. Pain, stiffness, and muscle strength, as measured by the Leppilahti score, were significantly better for the PRP group than the control group at 6 and 12 months post-surgery ($p < 0.05$). The PRP group also had significantly better general health, as estimated by the SF-36 score, than the control group at the 6-month time frame ($p < 0.05$). Ankle ROM was significantly greater for the PRP group than the control group at all time points

($p < 0.001$). For isokinetic calf strength, no significant difference was present between the PRP and control groups ($p > 0.05$).

The long-term effects of PRP in Achilles tendinopathy were assessed by Guelfi et al. [33], in which 83 tendons from 73 patients with chronic recalcitrant Achilles tendinopathies were treated with one PRP injection per tendon. Follow-ups were conducted at 3 weeks and 3- to 6-month intervals thereafter with the final follow-up occurring at a mean of 50.1 months post-injection. To evaluate healing, the Victorian Institute of Sport Assessment–Achilles (VISA-A) questionnaire and Blazina score were used, both to measure pain, function, and activity. At the final follow-up, VISA-A and Blazina scores significantly improved compared to baseline, indicating decreased pain and increased activity in the long term (VISA-A, $p < 0.01$; Blazina, $p < 0.05$). Patients rated 76 of the 83 tendons as satisfactory during follow-up sessions. The seven tendons that were rated below satisfactory were retreated with a second PRP injection occurring at a mean of 12 months after the first injection. All seven of these Achilles tendons proceeded to receive improved VISA-A and Blazina scores. Patients reported no tendon ruptures during the course of the study.

While PRP is safe and may provide some benefit, insufficient evidence exists that indicate its improved efficacy over other treatments for Achilles tendinopathy.

Patella Tendinopathy

Patellar tendinopathy, or "jumpers knee," is a degenerative disease that presents as chronic knee pain that often occurs due to overuse of the patellar tendon [34]. The underlying cause of this degenerative process is thought to occur due to a poor healing response of the tendon being subject to repetitive trauma. PRP preparations are being researched as an alternative non-operative treatment option with the potential to enhance tissue healing mechanisms in this injury (see Table 4.3).

In a double-blind, randomized controlled trial, Dagoo et al. [35] investigated leukocyte-rich PRP as a treatment for patellar tendinopathy. The study included 23 patients who failed conservative management; they were randomized into two groups (dry needling (DN) vs PRP injection) and followed for 26 weeks. Injections were guided using ultrasound. Patients were evaluated at 3, 6, 9, and 12 weeks and >6 month time points with the primary outcome being measured using the Victorian Institute of Sport Assessment (VISA) score for patellar tendinopathy. At 12 weeks following treatment, VISA scores improved by a mean ± standard deviation of 5.2 ± 12.5 points ($P = 0.20$) in the DN group ($n = 12$) and by 25.4 ± 23.2 points ($P = 0.01$) in the PRP group ($n = 9$); at ≥26 weeks, the scores improved by 33.2 ± 14.0 points ($P = 0.001$) in the DN group ($n = 9$) and by 28.9 ± 25.2 points ($P = 0.01$). The PRP group compared to the DN group had statistically significant improvement of symptoms at 12 weeks ($p = 0.02$) but no significant difference at 26 weeks ($p = 0.66$). Overall, this study found that ultrasound-guided leukocyte-rich PRP injections had accelerated recover compared with patients who received DN alone, but that results dissipated with time.

Table 4.3 PRP for patella tendinopathy

Study	Year	Study type	Control group	VISA-P score mean LR-PRP	P-value	VISA-P score mean control	P-value
Goesens et al. [37]	2012	PCS	Prior surgery, CSI, or ethoxysclerol therapy	Baseline = 39.1 +/− 16.6 18.4 months = 58.6 +/− 25.4	P = 0.003	Baseline = 41.8 +/− 14.3 18.4 months = 56.3 +/− 26.2	P = 0.093
Vetran et al. [26]	2013	RCT	ECSWT	Baseline = 55.3 +/− 14.3 12 months = 91.3 +/− 9.9	P = <0.0.5	Baseline = 56.1 +/− 19.9 12 months = 77.6 +/− 19.9	P = < 0.0.5
Dragoo et al. [25]	2014	RCT	Dry needling	Baseline = 47.4 +/− 14.3 12 week = 66.4 +/− 20.2 26 weeks = 67.8 +/− 21.9	– P = 0.02 P = 0.66	Baseline = 41 +/− 14.3 12 week = 52 +/− 20.3 26 weeks = 83.9 +/− 9.0	– – –

ECSWT extracorporeal shockwave therapy, *RCT* randomized controlled trial, *PCS* prospective cohort study, *LR-PRP* leukocyte-rich PRP

A randomized control trial by Vetrano et al. compared PRP to extracorporeal shockwave therapy (ECSWT) in treating patellar tendinopathy. Forty-six consecutive athletes with jumper's knee were selected for this study and randomized into 2 treatment groups: 2 autologous PRP injections over 2 weeks under ultrasound guidance (PRP group; $n = 23$) and 3 sessions of focused extracorporeal shockwave therapy (2.400 impulses at 0.17–0.25 mJ/mm^2 per session) (ECSWT group; $n = 23$). Patients were evaluated at 6- and 12-month follow-up using Victorian Institute of Sport Assessment–Patella (VISA-P) questionnaire, pain visual analog scale (VAS), and modified Blazina scale. The PRP group showed significantly better improvement than the ECSWT group in VISA-P, VAS scores at 6- and 12-month follow-up, and modified Blazina scale score at 12-month follow-up ($P < 0.05$ for all) [36].

A prospective cohort study by Goesens et al. evaluated outcome of patients with patellar tendinopathy treated with PRP and whether prior treatment affected efficacy of PRP injections. Thirty-six patients with patellar tendinopathy were selected and separated into two groups based on whether they had prior interventions. Patients were evaluated using Victorian Institute of Sport Assessment–Patellar (VISA-P) questionnaire, visual analogue scales (VAS), and pain with activities of daily living before and after PRP injections. They found that all patients regardless of prior intervention treated with PRP had statistically and clinically significant improvements in VAS scales ($p < 0.05$) [37]. Furthermore, patients that had not had previous treatments showed much larger improvements and had a much better potential for healing compared to patients who had received prior interventions.

Overall, research has demonstrated leukocyte-rich PRP injections are an effective treatment providing improved outcome scores for patients with patellar tendinopathy. However, there still remains insufficient evidence demonstrating its value over other treatment options to recommend it as part of standard treatment protocols for patellar tendinopathy.

Lateral Epicondylitis

Lateral epicondylitis (LE), commonly referred to as "tennis elbow," is an overuse injury leading to a degenerative process affecting the common extensor tendons at the lateral epicondyle of the distal humerus. It is the most common cause of elbow pain with a prevalence of about 1–3% in the general population with rates that have been described up to 10% in women [38]. The mechanism of injury most commonly reported in lateral epicondylitis is micro-tearing and degeneration due to repetitive trauma of the extensor carpi radialis brevis (ECRB) [39]. First-line treatment options are focused on non-operative interventions including bracing, cortisone injections, physical therapy, acupuncture, botulinum toxin injections, extracorporeal shockwave therapy (ECSWT), activity modification, and rest. Surgical procedures are often reserved for patients who have failed conservative therapies. Research on PRP in treating lateral epicondylitis has grown considerably given that no single treatment option has emerged as the best non-surgical therapy to reduce pain and improve function [40].

One of the difficulties in reviewing the literature on PRP for treating LE is the variability in preparations that are commercially available. In particular as it relates to LE, studies have found that leukocyte-rich PRP (LR-PRP) provides significant improvement in pain and function, whereas leukocyte-poor PRP (LP-PRP) results have been less conclusive. A randomized controlled trial by Palacio et al. [41] prospectively compared LP-PRP, CSI, and anesthetic injections for treating LE of the elbow. Sixty patients were randomly assigned one of the three treatment options and evaluated at 90 and 180 days using Disabilities of the Arm, Shoulder, and Hand (DASH) and Patient-Rated Tennis Elbow Evaluation (PRTEE) questionnaires. They concluded there was no evidence that one treatment was more effective than another in treating LE.

In contrast, in a prospective cohort study, Mishra et al. [42] evaluated the effectiveness of LR-PRP injection for patients who failed to respond to 3 months or more of conservative therapy for lateral epicondylitis. At 24 weeks, LR-PRP injection was associated with a statistically significant improvement in pain and residual elbow tenderness ($p = 0.009$) when compared to an injection of local anesthetic and needling [42].

When comparing LR-PRP to CSI for lateral epicondylitis, Peerbooms et al. [43] found that LR-PRP resulted in a greater reduction in pain and improved function compared to CSI. This was similarly demonstrated by Gosens et al. [44], where they found LR-PRP injections provided significantly greater symptomatic relief compared to CSI for LE. Furthermore, they found that while CSI did provide significant short-term relief, the effects diminished over time, returning to baseline after 26 weeks from time of injection.

Studies have demonstrated LR-PRP is a safe and effective treatment option for lateral epicondylitis, showing significant improvements in pain and function both short and long term when compared to controls (CSI, anesthetic, dry needling). However, questions remain regarding the cost-effectiveness, optimal preparation, and timing of intervention (see Table 4.4).

Table 4.4 PRP for lateral epicondylitis

Study	Year	Study type	Control group	Favors PRP?	PRP preparation	Level of evidence
Palacio et al. [41]	2016	RCT	Steroid	No	LP-PRP	I
Behera [45]	2015	RCT	Bupivacaine	Yes	LP-PRP	I
Mishra [42]	2013	RCT	Needling/ anesthetic	Yes	LR-PRP	II
Gosens et al. [44]	2011	RCT	Steroid	Yes	LR-PRP	–
Peerbooms et al. [43]	2010	RCT	Steroid	Yes	LR-PRP	–

RCT randomized controlled study, *LP-PRP* leukocyte-poor PRP, *LR-PRP* leukocyte-rich PRP

ACL Injury

Anterior cruciate ligament injuries make up 40–50% of ligamentous knee injuries, occurring predominantly during athletic activity [46]. Many people with ACL injury are recreational and elite athletes with significant incentive to have expeditious return to play following ACL injury and reconstruction. There are approximately 200,000 ACL reconstructions performed in the USA annually with a reported failure rate of 5–15% [47, 48]. PRP has gained interest in ACL surgery as an adjunct treatment with hope of growth factors promoting improved incorporation of ACL grafts [49].

A systematic review by Figueroa et al. examined studies comparing ACL graft maturation, integration time, and clinical outcomes with and without the use of PRP injection at the time of reconstruction. Although there was some heterogeneity among outcomes in the studies included in the analysis, they reported improved time to maturity if ACL graft. There was no difference regarding tunnel healing [50].

Studies examining the use of PRP for prevention of bony tunnel enlargement and graft-bone tunnel integration demonstrate conflicting results. Compelling evidence in favor of PRP was found in double-blinded RCT which examined the use of a series of four intra-articular injections of PRP following ACL reconstruction and utilized CT scan to measure bone tunnel width at 1, 6, and 12 months postoperative. The group treated with PRP was found to have less tunnel widening at 6 and 12 months compared to saline injection controls. Furthermore, WOMAC stiffness subscale was found to be consistently better in those treated with PRP 1-year postoperative [51]. However, these results have failed to be consistently demonstrated in the literature [49, 50].

Although promising results exist regarding graft maturation, no studies have clearly demonstrated its clinical significance. Furthermore, conflicting evidence are noted regarding the use of PRP to decrease ACL graft tunnel widening and expedite healing [49, 50]. The use of PRP in ACL reconstruction is a decision that should be made with the patient after a thorough discussion regarding its possible but not consistently proven benefits.

Meniscus Injury

The meniscus comprises the medial and lateral meniscus, two C-shaped fibrocartilaginous structures attached anteriorly and posteriorly to the tibial plateau. The meniscus provides secondary stabilization, load distribution, load sharing, lubrication, and proprioception to the knee joint [52]. Meniscus injury is one of the most common musculoskeletal injuries in athletes with an overall incidence rate reported at 8.27% [53]. Studies have shown that the compartment pressures within the knee increase significantly with meniscus tearing and can further increase after partial or total meniscectomy but are improved with meniscus repair [54, 55]. The challenge the orthopedic surgeons face when repairing the meniscus is that only the peripheral 25–30% of the meniscus is vascularized [56]. Therefore, healing potential of more centralized tears is greatly limited. The biomechanical implications of meniscus injury in combination with the biologic challenge of a lack of vascularity create great interest in the use of PRP as a means of treatment.

Ishida et al. reported the effects of PRP on meniscal tissue regeneration, both in vitro and in vivo. In the in vitro study, PRP stimulated DNA synthesis, ECM synthesis, and mRNA expression of biglycan and decorin in monolayer meniscal cell cultures [57]. In the in vivo study, full-thickness defects were produced in the avascular region of the rabbit meniscus. Gelatin hydrogel was used to deliver PRP into the defects. Histologic analysis of the healing tissue 12 weeks after surgery showed significantly better meniscal repair in animals that received PRP with gelatin hydrogel compared with animals in the control group.

Clinically, Griffin et al. [58] reported a comparison between 11 isolated meniscus repairs that were augmented with PRP and 15 isolated repairs that were performed without PRP at a minimum follow-up of 2 years. No difference was noted in clinical score, rate of return to sports, and rate of reoperation. These results are in contrast to those by Pujol et al., who examined outcomes of repair in patients with symptomatic horizontal meniscus tears extending into the avascular zone with or without PRP injection. They found an improvement in KOOS pain and returned to sport ($p < 0.05$) in the PRP injection group [59].

Ultimately, no definitive recommendations can be made at this time regarding PRP injection in the treatment of meniscus tears.

Rotator Cuff Injury

Treatment for rotator cuff tears, including massive tears and chronic injuries, is troubled by high failure rates as well as high rates of retear. Tendon retears may be due to a specific reinjury at the repair site but also may reflect incomplete or failed primary healing after surgery. This has led to an increased interest in evaluating the benefits of orthobiologics, in particular PRP, in potentially improving the healing process and decreasing rates of failure (see Table 4.5).

Contradictory results exist for the benefits of PRP in rotator cuff injuries. PRP does contain factors that improve tendon-bone healing such as TGF-β1 and IGF-1. These factors may contribute to improvements in tendinopathy healing similar to factors that improve tendon regeneration by decreasing inflammation. However, despite advances in implant technology and repair techniques, a poor intrinsic healing environment at the bone interface results in relatively high rates of nonhealing and retears. While some studies have demonstrated a positive effect using LP-PRP in setting of rotator cuff injuries, there remains conflicting level 1 studies that have failed to produce similar results.

In a randomized controlled trial, Randell et al. [60] in 2011 evaluated use of PRP in arthroscopic rotator cuff repair with 2-year follow-up. They reported initial short-term improvements in the PRP-treated groups; however, these benefits dissipated over time with no difference between PRP-treated group and control using UCLA scores at 6, 12, or 24 months. Similarly, Castricini et al. [61] performed a RCT

Table 4.5 PRP for rotator cuff injury

Study (level of evidence)	Procedure and disease	PRP preparation method (kit manufacturer)	Application	Activation method	Combined treatments	Comparison group	Outcome
Randell et al. [60] (I)	Arthroscopic rotator cuff repair	Double centrifugation (GPSII-Plasmax-platelet concentration system; Biomet biologics, Warsaw, IN, USA)	Surgical	10% calcium chloride and bovine thrombin solution	Arthroscopic repair and PRP	Yes (arthroscopic repair and no PRP)	PRP reduced pain in the first postoperative months The long-term results of subgroups of grade 1 and 2 tears suggest that PRP positively affected cuff rotator healing
Castricini et al. [61] (I)	Arthroscopic rotator cuff repair	Cascade autologous platelet system (MTF [Musculoskeletal Transplant Foundation], Edison, New Jersey) + platelet-rich fibrin matrix (PRFM)	Surgical	No	Arthroscopic repair and autologous PRFM	Yes (arthroscopic repair and no PRFM)	No significant benefit in rotator cuff healing with addition of PRFM for small- and medium-sized rotator cuff tears
Wang et al. [62] (I)	Arthroscopic rotator cuff repair	(Arthrex autologous conditioned plasma (ACP) system, Naples, FL)	Surgical	10% calcium chloride	Arthroscopic repair and PRP	Yes (arthroscopic repair and no PRP)	No significant improvement in early tendon-bone healing or functional recover with addition of PRP following arthroscopic rotator cuff repair
Malavolta et al. [64] (I)	Arthroscopic rotator cuff repair	Double centrifugation (Haemonetics MCS1 9000 blood cell separator and 994-CFE apheresis set (Haemonetics Corp, Braintree, MA))	Surgical	10% calcium chloride and bovine thrombin solution	Arthroscopic repair and PRP	Yes (arthroscopic repair and no PRP)	No significant improvement in clinical results at 24-month follow-up with addition of PRP following arthroscopic rotator cuff repair. No change in retear rates with addition of PRP

(continued)

Table 4.5 (continued)

Study (level of evidence)	Procedure and disease	PRP preparation method (kit manufacturer)	Application	Activation method	Combined treatments	Comparison group	Outcome
D'Ambrosi et al [66] (I)	Arthroscopic rotator cuff repair	Double centrifugation (GPSIII-plasmax-platelet concentration system; Biomet biologics, Warsaw, IN, USA)	Surgical	10% calcium chloride and bovine thrombin solution	Arthroscopic repair and PRP	Yes (arthroscopic repair and no PRP)	PRP leads to a reduction in pain during a short-term follow-up (6 months). No difference in tendon healing or re-rupture rates with addition of PRP compared to control
Holtby et al. [67]	Arthroscopic rotator cuff repair	(SmartPrep 2 system, harvest autologous hemobiologics)	Surgical	No	Arthroscopic repair and PRP	Yes (arthroscopic repair and no PRP)	PRP leads to significant reduction in short-term pain following repair of small- to medium-sized rotator cuff tears without any significant impact on patient-oriented outcome measures or structural integrity of the repair compared with control group

evaluating PRP augmentation for arthroscopic rotator cuff repairs for small- and medium-sized tears. They reported no significant difference in outcome score or tendon healing in both small and medium size tears treated with PRP. In 2015, Wang et al. [62] assessed use of PRP in setting of rotator cuff repairs and compared results with both patient-reported outcomes and MRI evaluation. They reported no significant difference in constant and tendon scores on MRIs with the use of PRP in setting of rotator cuff repairs. A meta-analysis performed by Cai et al. [63] in 2015 contained five qualifying level 1 studies. The review concluded no overall difference in outcomes in patients who underwent rotator cuff repair augmented with PRP. Overall results have been disappointing regarding PRPs benefit in aiding rotator cuff repair healing.

Other studies have evaluated PRPs potential ability to decrease retears following rotator cuff repairs with conflicting results. In 2014, Malavolta et al. [64] performed randomized control study assessing effect of PRP in patients undergoing rotator cuff repair of small or medium size tears. A total of 54 patients were included with an equal number of patients being randomly assigned into both groups, surgery plus PRP group versus surgery alone. They concluded that PRP did not enhance healing or reduce the chance of retears compared to controls. A meta-analysis performed in 2015 by Warth et al. [65] containing eight level 1 and level 2 studies assesses PRP on rotator cuff repairs. They concluded there was no difference in the outcomes for patients for up to 24 months. However, they found that when PRP is applied to the tendon bone interface, it produced better outcomes than liquid PRP injected over the tendon. Furthermore, they concluded there was a decreased rate of retear with PRP for medium to large tears.

Finally, limited evidence exists that PRP may be beneficial for postoperative pain following rotator cuff repair. A study by D'Ambroisi et al. [66] evaluated efficacy of PRP in setting of rotator cuff repair. They conclude that patients who received PRP injections at time of surgery had significant reduction in postoperative pain compared to controls in the short term; however, results were similar at longer follow-up. Similar results were obtained by Holtby et al. [67] in which they found that PRP has a short-term effect on perioperative pain without any significant impact on patient-oriented outcome measures or structural integrity of the repair compared with control groups. Overall, PRP has been proven to be a safe adjunct that may reduce perioperative pain.

Similar to other aspects of orthopedic care, variations in PRP preparations and technique often lead to confounding results. Given the contradictory results in the literature, no strong recommendation can be made to support routine use of PRP preparations in setting of rotator cuff tears. More high-quality studies are needed to assess PRPs true value in the setting of rotator cuff tendon healing and repair.

Conclusion

PRP is gaining popularity as an orthobiologic, showing increasing promise in treating many orthopedic injuries. Moreover, PRP injections are low cost and less invasive than surgical alternatives, making them accessible and low risk [68]. Evidence of PRP's effectiveness continues to grow, yet the criteria needed to create the optimal PRP preparation still remains inconclusive. More level 1 clinical studies and meta-analyses are needed to compare the efficacies of different PRP preparations on various orthopedic pathologies [1].

References

1. Navani A, Li G, Chrystal J. Platelet rich plasma in musculoskeletal pathology: a necessary rescue or a lost cause? Pain Physician. 2017;20(3):E345–56.
2. Wojdasiewicz P, Poniatowski LA, Szukiewicz D. The role of inflammatory and anti-inflammatory cytokines in the pathogenesis of osteoarthritis. Mediat Inflamm. 2014;2014: 561459.
3. Xie X, Zhang C, Tuan RS. Biology of platelet-rich plasma and its clinical application in cartilage repair. Arthritis Res Ther. 2014;16(1):204.
4. Russell RP, Apostolakos J, Hirose T, Cote MP, Mazzocca AD. Variability of platelet-rich plasma preparations. Sports Med Arthrosc Rev. 2013;21(4):186–90.
5. Dhurat R, Sukesh M. Principles and methods of preparation of platelet-rich plasma: a review and author's perspective. J Cutan Aesthet Surg. 2014;7(4):189–97.
6. Beitzel K, Allen D, Apostolakos J, et al. US definitions, current use, and FDA stance on use of platelet-rich plasma in sports medicine. J Knee Surg. 2015;28(1):29–34.
7. Gu W, Li T, Shi Z, et al. Management of Hepple stage V osteochondral lesion of the talus with a platelet-rich plasma scaffold. Biomed Res Int. 2017;2017:6525373.
8. Gormeli G, Karakaplan M, Gormeli CA, Sarikaya B, Elmali N, Ersoy Y. Clinical effects of platelet-rich plasma and hyaluronic acid as an additional therapy for talar osteochondral lesions treated with microfracture surgery: a prospective randomized clinical trial. Foot Ankle Int. 2015;36(8):891–900.
9. Wei LC, Lei GH, Sheng PY, et al. Efficacy of platelet-rich plasma combined with allograft bone in the management of displaced intra-articular calcaneal fractures: a prospective cohort study. J Orthop Res. 2012;30(10):1570–6.
10. Repetto I, Biti B, Cerruti P, Trentini R, Felli L. Conservative treatment of ankle osteoarthritis: can platelet-rich plasma effectively postpone surgery? J Foot Ankle Surg. 2017;56(2): 362–5.
11. Fukawa T, Yamaguchi S, Akatsu Y, Yamamoto Y, Akagi R, Sasho T. Safety and efficacy of intra-articular injection of platelet-rich plasma in patients with ankle osteoarthritis. Foot Ankle Int. 2017;38(6):596–604.
12. Schwartz A. A promising treatment for athletes, in blood. The New York Times. 2009.
13. World Anti-Doping Agency. World Anti-Doping Agency List of prohibited substances and methods. 2020. https://www.wada-ama.org/en/content/what-is-prohibited/prohibited-at-all-times. Published 2020. Accessed December 1, 2020.
14. Mishra A, Harmon K, Woodall J, Vieira A. Sports medicine applications of platelet rich plasma. Curr Pharm Biotechnol. 2012;13(7):1185–95.
15. Ekstrand J, Hagglund M, Walden M. Epidemiology of muscle injuries in professional football (soccer). Am J Sports Med. 2011;39(6):1226–32.
16. Ahmad CS, Redler LH, Ciccotti MG, Maffulli N, Longo UG, Bradley J. Evaluation and management of hamstring injuries. Am J Sports Med. 2013;41(12):2933–47.
17. Orchard J, Best TM, Verrall GM. Return to play following muscle strains. Clin J Sport Med. 2005;15(6):436–41.
18. Heiderscheit BC, Sherry MA, Silder A, Chumanov ES, Thelen DG. Hamstring strain injuries: recommendations for diagnosis, rehabilitation, and injury prevention. J Orthop Sports Phys Ther. 2010;40(2):67–81.
19. Hamilton B. Hamstring muscle strain injuries: what can we learn from history? Br J Sports Med. 2012;46(13):900–3.

20. Orchard JW, Best TM, Mueller-Wohlfahrt HW, et al. The early management of muscle strains in the elite athlete: best practice in a world with a limited evidence basis. Br J Sports Med. 2008;42(3):158–9.
21. Laumonier T, Menetrey J. Muscle injuries and strategies for improving their repair. J Exp Orthop. 2016;3(1):15.
22. Menetrey J, Kasemkijwattana C, Day CS, et al. Growth factors improve muscle healing in vivo. J Bone Joint Surg Br. 2000;82(1):131–7.
23. Grassi A, Napoli F, Romandini I, et al. Is platelet-rich plasma (PRP) effective in the treatment of acute muscle injuries? A systematic review and meta-analysis. Sports Med. 2018;48(4):971–89.
24. Quarteiro ML, Tognini JR, de Oliveira EL, Silveira I. The effect of platelet-rich plasma on the repair of muscle injuries in rats. Rev Bras Ortop. 2015;50(5):586–95.
25. Hammond JW, Hinton RY, Curl LA, Muriel JM, Lovering RM. Use of autologous platelet-rich plasma to treat muscle strain injuries. Am J Sports Med. 2009;37(6):1135–42.
26. Hamid MSA, Mohamed Ali MR, Yusof A, George J, Lee LP. Platelet-rich plasma injections for the treatment of hamstring injuries: a randomized controlled trial. Am J Sports Med. 2014;42(10):2410–8.
27. Rossi LA, Molina Romoli AR, Bertona Altieri BA, Burgos Flor JA, Scordo WE, Elizondo CM. Does platelet-rich plasma decrease time to return to sports in acute muscle tear? A randomized controlled trial. Knee Surg Sports Traumatol Arthrosc. 2017;25(10):3319–25.
28. Pas HI, Reurink G, Tol JL, Weir A, Winters M, Moen MH. Efficacy of rehabilitation (lengthening) exercises, platelet-rich plasma injections, and other conservative interventions in acute hamstring injuries: an updated systematic review and meta-analysis. Br J Sports Med. 2015;49(18):1197–205.
29. Reurink G, Goudswaard GJ, Moen MH, et al. Rationale, secondary outcome scores and 1-year follow-up of a randomised trial of platelet-rich plasma injections in acute hamstring muscle injury: the Dutch Hamstring Injection Therapy study. Br J Sports Med. 2015;49(18):1206–12.
30. Hamilton B, Tol JL, Almusa E, et al. Platelet-rich plasma does not enhance return to play in hamstring injuries: a randomised controlled trial. Br J Sports Med. 2015;49(14):943–50.
31. Martinez-Zapata MJ, Orozco L, Balius R, et al. Efficacy of autologous platelet-rich plasma for the treatment of muscle rupture with haematoma: a multicentre, randomised, double-blind, placebo-controlled clinical trial. Blood Transfus. 2016;14(2):245–54.
32. Zou J, Mo X, Shi Z, et al. A prospective study of platelet-rich plasma as biological augmentation for acute Achilles tendon rupture repair. Biomed Res Int. 2016;2016:9364170.
33. Guelfi M, Pantalone A, Vanni D, Abate M, Guelfi MG, Salini V. Long-term beneficial effects of platelet-rich plasma for non-insertional Achilles tendinopathy. Foot Ankle Surg. 2015;21(3):178–81.
34. Kia C, Baldino J, Bell R, Ramji A, Uyeki C, Mazzocca A. Platelet-rich plasma: review of current literature on its use for tendon and ligament pathology. Curr Rev Musculoskelet Med. 2018;11(4):566–72.
35. Dragoo JL, Wasterlain AS, Braun HJ, Nead KT. Platelet-rich plasma as a treatment for patellar tendinopathy: a double-blind, randomized controlled trial. Am J Sports Med. 2014;42(3):610–8.
36. Vetrano M, Castorina A, Vulpiani MC, Baldini R, Pavan A, Ferretti A. Platelet-rich plasma versus focused shock waves in the treatment of jumper's knee in athletes. Am J Sports Med. 2013;41(4):795–803.
37. Gosens T, Den Oudsten BL, Fievez E, van't Spijker P, Fievez A. Pain and activity levels before and after platelet-rich plasma injection treatment of patellar tendinopathy: a prospective cohort study and the influence of previous treatments. Int Orthop. 2012;36(9):1941–6.
38. Verhaar JA. Tennis elbow. Anatomical, epidemiological and therapeutic aspects. Int Orthop. 1994;18(5):263–7.
39. Arirachakaran A, Sukthuayat A, Sisayanarane T, Laoratanavoraphong S, Kanchanatawan W, Kongtharvonskul J. Platelet-rich plasma versus autologous blood versus steroid injection in lateral epicondylitis: systematic review and network meta-analysis. J Orthop Traumatol. 2016;17(2):101–12.

40. Kwapisz A, Prabhakar S, Compagnoni R, Sibilska A, Randelli P. Platelet-rich plasma for elbow pathologies: a descriptive review of current literature. Curr Rev Musculoskelet Med. 2018;11(4):598–606.
41. Palacio EP, Schiavetti RR, Kanematsu M, Ikeda TM, Mizobuchi RR, Galbiatti JA. Effects of platelet-rich plasma on lateral epicondylitis of the elbow: prospective randomized controlled trial. Rev Bras Ortop. 2016;51(1):90–5.
42. Mishra AK, Skrepnik NV, Edwards SG, et al. Efficacy of platelet-rich plasma for chronic tennis elbow: a double-blind, prospective, multicenter, randomized controlled trial of 230 patients. Am J Sports Med. 2014;42(2):463–71.
43. Peerbooms JC, Sluimer J, Bruijn DJ, Gosens T. Positive effect of an autologous platelet concentrate in lateral epicondylitis in a double-blind randomized controlled trial: platelet-rich plasma versus corticosteroid injection with a 1-year follow-up. Am J Sports Med. 2010;38(2):255–62.
44. Gosens T, Peerbooms JC, van Laar W, den Oudsten BL. Ongoing positive effect of platelet-rich plasma versus corticosteroid injection in lateral epicondylitis: a double-blind randomized controlled trial with 2-year follow-up. Am J Sports Med. 2011;39(6):1200–8.
45. Behera P, Dhillon M, Aggarwal S, Marwaha N, Prakash M. Leukocyte-poor platelet-rich plasma versus bupivacaine for recalcitrant lateral epicondylar tendinopathy. J Orthop Surg (Hong Kong). 2015;23(1):6–10.
46. Miller MD. Anterior cruciate ligament. Clin Sports Med. 2017;36(1):xiii–xiv.
47. Yabroudi MA, Bjornsson H, Lynch AD, et al. Predictors of revision surgery after primary anterior cruciate ligament reconstruction. Orthop J Sports Med. 2016;4(9):2325967116666039.
48. Kamath GV, Redfern JC, Greis PE, Burks RT. Revision anterior cruciate ligament reconstruction. Am J Sports Med. 2011;39(1):199–217.
49. Andriolo L, Di Matteo B, Kon E, Filardo G, Venieri G, Marcacci M. PRP augmentation for ACL reconstruction. Biomed Res Int. 2015;2015:371746.
50. Figueroa D, Figueroa F, Calvo R, Vaisman A, Ahumada X, Arellano S. Platelet-rich plasma use in anterior cruciate ligament surgery: systematic review of the literature. Arthroscopy. 2015;31(5):981–8.
51. Darabos N, Haspl M, Moser C, Darabos A, Bartolek D, Groenemeyer D. Intraarticular application of autologous conditioned serum (ACS) reduces bone tunnel widening after ACL reconstructive surgery in a randomized controlled trial. Knee Surg Sports Traumatol Arthrosc. 2011;19(Suppl 1):S36–46.
52. McDermott ID, Amis AA. The consequences of meniscectomy. J Bone Joint Surg Br. 2006;88(12):1549–56.
53. Jones JC, Burks R, Owens BD, Sturdivant RX, Svoboda SJ, Cameron KL. Incidence and risk factors associated with meniscal injuries among active-duty US military service members. J Athl Train. 2012;47(1):67–73.
54. Zhang AL, Miller SL, Coughlin DG, Lotz JC, Feeley BT. Tibiofemoral contact pressures in radial tears of the meniscus treated with all-inside repair, inside-out repair and partial meniscectomy. Knee. 2015;22(5):400–4.
55. Bedi A, Kelly NH, Baad M, et al. Dynamic contact mechanics of the medial meniscus as a function of radial tear, repair, and partial meniscectomy. J Bone Joint Surg Am. 2010;92(6):1398–408.
56. Arnoczky SP, Warren RF. Microvasculature of the human meniscus. Am J Sports Med. 1982;10(2):90–5.
57. Ishida K, Kuroda R, Miwa M, et al. The regenerative effects of platelet-rich plasma on meniscal cells in vitro and its in vivo application with biodegradable gelatin hydrogel. Tissue Eng. 2007;13(5):1103–12.
58. Griffin JW, Hadeed MM, Werner BC, Diduch DR, Carson EW, Miller MD. Platelet-rich plasma in meniscal repair: does augmentation improve surgical outcomes? Clin Orthop Relat Res. 2015;473(5):1665–72.
59. Pujol N, Salle De Chou E, Boisrenoult P, Beaufils P. Platelet-rich plasma for open meniscal repair in young patients: any benefit? Knee Surg Sports Traumatol Arthrosc. 2015;23(1):51–8.

60. Randelli P, Arrigoni P, Ragone V, Aliprandi A, Cabitza P. Platelet rich plasma in arthroscopic rotator cuff repair: a prospective RCT study, 2-year follow-up. J Shoulder Elb Surg. 2011;20(4):518–28.
61. Castricini R, Longo UG, De Benedetto M, et al. Platelet-rich plasma augmentation for arthroscopic rotator cuff repair: a randomized controlled trial. Am J Sports Med. 2011;39(2):258–65.
62. Wang A, McCann P, Colliver J, et al. Do postoperative platelet-rich plasma injections accelerate early tendon healing and functional recovery after arthroscopic supraspinatus repair? A randomized controlled trial. Am J Sports Med. 2015;43(6):1430–7.
63. Cai YZ, Zhang C, Lin XJ. Efficacy of platelet-rich plasma in arthroscopic repair of full-thickness rotator cuff tears: a meta-analysis. J Shoulder Elb Surg. 2015;24(12):1852–9.
64. Malavolta EA, Gracitelli ME, Ferreira Neto AA, Assuncao JH, Bordalo-Rodrigues M, de Camargo OP. Platelet-rich plasma in rotator cuff repair: a prospective randomized study. Am J Sports Med. 2014;42(10):2446–54.
65. Warth RJ, Dornan GJ, James EW, Horan MP, Millett PJ. Clinical and structural outcomes after arthroscopic repair of full-thickness rotator cuff tears with and without platelet-rich product supplementation: a meta-analysis and meta-regression. Arthroscopy. 2015;31(2):306–20.
66. D'Ambrosi R, Palumbo F, Paronzini A, Ragone V, Facchini RM. Platelet-rich plasma supplementation in arthroscopic repair of full-thickness rotator cuff tears: a randomized clinical trial. Musculoskelet Surg. 2016;100(Suppl 1):25–32.
67. Holtby R, Christakis M, Maman E, et al. Impact of platelet-rich plasma on arthroscopic repair of small- to medium-sized rotator cuff tears: a randomized controlled trial. Orthop J Sports Med. 2016;4(9):2325967116665595.
68. Cavallo C, Roffi A, Grigolo B, et al. Platelet-rich plasma: the choice of activation method affects the release of bioactive molecules. Biomed Res Int. 2016;2016:6591717.

PRP for the Treatment of Osteoarthritis Pain

5

Lakshmi S. Nair

Abbreviations

ADAMTS:	A disintegrin-like and metalloproteinase with thrombospondin motifs
ADP:	Adenosine diphosphate
ATP:	Adenosine triphosphate
BMScs:	Bone marrow-derived stem cells
Cox-2:	Cyclooxygenase 2
CXCR$_4$:	C-X-C chemokine receptor type
DAMPs:	Disease-activating molecular patterns
FGF:	Fibroblast growth factor
HA:	Hyaluronic acid
HGF:	Hepatocyte growth factor
IGF:	Insulin-like growth factor
IGF-1:	Insulin-like growth factor-1
IKBα:	Inhibitor of KBα
IKDC:	International Knee Documentation Committee
IL-1R:	Interleukin 1 receptor
IL-1Ra:	Interleukin 1 receptor antagonist
IL-4:	Interleukin-4

L. S. Nair (✉)
Department of Orthopaedic Surgery, University of Connecticut Health Center, Farmington, CT, USA

The Connecticut Convergence Institute for Translation in Regenerative Engineering, University of Connecticut Health Center, Farmington, CT, USA

Department of Materials Science and Engineering, Institute of Material Science, University of Connecticut, Storrs, CT, USA

Department of Biomedical Engineering, University of Connecticut, Storrs, CT, USA
e-mail: nair@uchc.edu

LP-PRP: Leukocyte-poor PRP
LR-PRP: Leukocyte-rich PRP
MMP: Matrix metalloprotease
NFkB: Nuclear factor kB
NO: Nitric oxide
OA: Osteoarthritis
PDGF: Platelet-derived growth factor
PGE2: Prostaglandin E2
PRP: Platelet-rich plasma
PRPr: Platelet-rich plasma releasate
RCT: Randomized clinical trial
TGF-β: Transforming growth factor-beta
TNF-α: Tumor necrosis factor alpha
VAS: A scale that intuitively quantifies pain in the knee. A lower score indicates milder pain
WOMAC: Western Ontario and McMaster Universities Arthritis Index. Rating scale for assessing the structure and function of the knee joint in terms of pain, stiffness, and joint function. A lower score indicates better function

Introduction

Over the past two decades, autologous blood products such as platelet-rich plasma (PRP) have been extensively used in various clinical areas such as orthopedic and spinal surgery, oral, periodontal, cosmetic, and maxillofacial surgery. Platelet-rich plasma (PRP) is a term originated in the 1970s to describe the plasma with a platelet count above that of peripheral blood [1]. Platelets are anucleated and discoid cells derived from megakaryocytes. The primary role of platelets is known to be hemostasis to maintain the integrity of vasculature upon injury. Recent studies demonstrated a more expanded functional role of platelets in modulating inflammation, angiogenesis, wound healing, and cancer metastasis [2]. By now, it is evident that platelets modulate these biological functions mostly through a variety of bioactive molecules present in platelet secretomes. More than 300 bioactive molecules have been identified in platelet secretome, and these molecules are released from platelet secretory granules. Three principle types of secretary granules are identified in platelets, and they are α-granules, dense granules, and lysosomes. α-Granules are the most abundant of the platelet granules and contain membrane proteins, adhesive molecules such as von Willebrand factor and fibrinogen, coagulants, various chemokines, cytokines, and growth factors [3]. Dense granules which are second in abundance contain small molecules such as adenosine diphosphate (ADP), adenosine triphosphate (ATP), serotonin, epinephrine, calcium, pyrophosphate, and polyphosphate. Platelet lysosomes, on the other hand, contain acid hydrolases such as cathepsins, hexosaminidase, β-galactosidase, arylsulfatase, β-glucuronidase, and acid phosphatase. In short, with these multiple secretary granules, ribosomes, signal transduction pathways, and multiple receptors, platelets serve as unique secretory machines. Even though it is not clearly understood how platelets regulate the secretory processes, the presence of molecules with apparent antagonistic functions suggests differential secretory kinetics and activation-dependent synthesis and release of cytokines from platelets [4]. The differential pattern or composition of platelet secretion may be potentially controlled by the presence of different platelet agonists at the sites of tissue injury.

Since the 1980s, platelet transfusion is commonly performed in the United States [5]. Several methods are currently available for preparing PRP from whole blood, and consequently, the composition and the activation state of PRP can significantly vary depending on the preparation process. Ehrenfest et al. have classified the methods of PRP preparation into four main families depending on the platelet count, presence, or absence of leucocytes and fibrin architecture [6]. These are (1) leucocyte-poor PRP products (without leucocytes and with a low-density fibrin network upon activation); (2) PRP with leucocytes (with a low-density fibrin network upon activation); (3) leucocyte-poor platelet-rich fibrin preparation (without leucocytes and with a high-density fibrin network); and (4) leucocyte and platelet-rich fibrin (with leucocytes and high-density fibrin network). Magalon et al. proposed the DEPA classification based on four components including the platelet activation process which can significantly influence the composition of the bioactive factors [7]. The components of DEPA classification are (1) dose of injected platelets; (2) efficiency of the production (percentage of the platelets recovered in the PRP from the blood); (3) purity of the PRP (relative composition of the platelets, leucocytes, and RBCs); and (4) the activation process (whether exogenous clotting factor was used to activate platelets). In short, studies so far have identified major variables that may affect the efficacy of PRP formulations such as PRP composition, PRP activation mode, and the method of PRP application [8]. Lack of standardization of PRP preparations, methods of PRP delivery, differing techniques of platelet activation, and procedural heterogeneity complicates critical evaluation of the clinical efficacy of PRP.

Platelets play a multifaceted role in promoting healing by supporting blood coagulation, modulating immune response, and promoting mesenchymal cell migration, as well as by regulating cellular apoptosis. The natural tissue healing potential of PRP has been attributed to the variety of autologous bioactive proteins present in it such as immune mediators, chemokines, growth factors, enzymes, and their inhibitors. Particularly, the presence of various growth factors in PRP such as platelet-derived growth factor (PDGF), transforming growth factor-beta (TGF-β), fibroblast growth factor (FGF), hepatocyte growth factor (HGF), insulin-like growth factor-1 (IGF-1), vascular endothelial growth factor (VEGF), epidermal growth factor (EGF), and adhesion molecules (e.g., fibrin, fibronectin, and vitronectin) which are involved in the wound healing process makes PRP an attractive candidate for widespread clinical use. Moreover, the autologous source of PRP makes it highly attractive for clinical use as it minimizes potential side effects and disease transmission. The use of PRP technology and therapies for orthopedic applications has been focused to improve bone and soft tissue regeneration and as an adjunct in surgical reconstruction process [9]. In addition to these, recent studies showed the exciting capabilities of PRP injections to decrease pain in patients suffering from musculo-skeletal conditions such as osteoarthritis. This chapter presents an overview of the potential of intra-articular injection of PRP for treating osteoarthritic pain.

PRP for Treating Osteoarthritis Pain

Osteoarthritis (OA) is a degenerative disease that affects the knee and hip joints in humans and involves joint damage and progressive deterioration of the joint archi-tecture. The degenerative process has been attributed to the imbalance of anabolic and catabolic activities in the cartilage resulting in eventual cartilage loss. The hall-mark symptom of OA is joint pain. The pain and the associated physical disability of osteoarthritis have shown to significantly affect the quality of life of people. Of all the joints, knee joint is most commonly affected by OA and is the leading cause of disability. The prevalence of OA is increasing in the United States, and a recent epidemiological survey showed that incidence of knee OA in the United States has doubled since the mid-twentieth century [10]. It is estimated that by 2030 the num-ber of people with OA may reach ~70 million [11].

A wide range of treatments are currently in use to reduce OA-related pain and functional disability including non-pharmacological interventions to dietary supple-ments, pharmacological therapies, minimally invasive intra-articular injections, and finally invasive surgical approaches [12, 13]. Osteoarthritis management via non-surgical or minimally invasive procedures focus to reduce pain and inflammation, improve function, and reduce disability. Non-drug treatments for OA include exer-cise and physical therapy. Topical or oral nonsteroidal anti-inflammatory drugs are still considered as promising first-line analgesic medication. However, the reported cardiovascular, gastrointestinal, and renal side effects of systemic corticosteroids greatly limit their use for treating knee OA [14]. Even though opioids were pre-scribed for treating OA pain, they have shown only modest effects to alleviate pain with significant side effects including addiction [15]. Minimally invasive intra-articular therapies present itself as an attractive alternative as it has several physio-logical and practical advantages such as improved safety over systemic medications particularly when comorbidities are present and improved bioavailability. In spite of this, intra-articular therapies are currently considered only as the last non-operative therapy option with a goal to prolong the need for operative intervention. Intra-articular corticosteroid injections combined with local anesthetics are frequently performed to treat knee joint pain. Recent studies show potential complications associated with intra-articular corticosteroid injections which necessitate the need to find alternative approaches [16]. Viscosupplementation is another commonly practiced approach which involves a series of intra-articular injection of hyaluronic acid (HA) [17]. The mechanism of action of HA upon intra-articular injection has not been well established. The favorable effects of HA have been attributed to its potential to increase the viscoelasticity of the synovial fluid as well as its favorable interactions with chondrocytes. Even though the clinical results are still inconclusive, intra-articular HA injection has shown to have short-term effects in decreasing pain in patients with early to moderate knee OA [18]. Novel and/or alternative therapeutic strategies are being investigated to improve intra-articular treatment of OA to reduce joint pain and slow down the progression of cartilage degeneration. Due to the ability of PRP to accelerate wound healing process and cellular proliferation as well as enhance chondrogenic differentiation, there is an increasing interest in intra-articular PRP injection for treating OA. Recent years, therefore, saw a significant increase in the clinical use of intra-articular delivery of PRP for degenerative OA due to its safety, ease of production, and administration [19]. Following sections review the recent preclinical and clinical data evaluating the efficacy of intra-articular PRP therapy in reducing OA-related joint pain and the currently known mechanism of action of PRP upon intra-articular injection.

Mechanism of Action of PRP upon Intra-Articular Injection

Although the exact mechanism of action of intra-articular PRP is not yet clearly elucidated, preclinical and clinical studies have shown the potential of intra-articular PRP injections in decreasing OA pain. The mechanisms involved in reducing joint pain upon intra-articular injection PRP may include direct inhibition of tissue nociceptors, inhibition/reduction of joint inflammation, stimulation of endogenous HA, and reduction of catabolic activities via inhibition of MMPs.

Durant et al. investigated the protective nature of PRP against chondrocyte death when exposed to corticosteroids or local anesthetics. The addition of PRP significantly increased chondrocyte viability and proliferation in the presence of local anesthetics or corticosteroid demonstrating its cytoprotective effect [20]. The anabolic and chondrogenic activities of PRP have been attributed to the presence of IGF-1 and TGF-β, as both these growth factors are known to have beneficial effects in modulating cell proliferation and ECM deposition. This was supported by a study which investigated the effect of PRP releasate (PRPr) on human articular chondrocyte [21]. PRPr was prepared by two cycles of freezing and thawing followed by a final centrifugation. The study demonstrated that PRPr can significantly enhance the mitogenic and differentiating properties of human chondrocytes in mono and three-dimensional cultures. The anabolic effect of PRP releasate on bone marrow-derived stem cells (BMScs) is shown in a study by Zaky et al., wherein platelet lysate significantly enhanced human BMSc proliferation and chondrogenic differentiation [22]. The study concluded the potential of platelet lysate as a fetal bovine serum substitute to support BMSc chondrogenic differentiation. A study by Drengk et al. corroborated the above results showing the proliferative and chondrogenic effect of PRP on mesenchymal stem cell [23]. However, unlike the previous studies, this study showed that exposure to PRP can led to chondrocyte proliferation and dedifferentiation with decrease in chondrogenic phenotype. The contradicting preclinical and clinical efficacy can be attributed to a lack of standardized PRP preparation since the concentration and the activation state of the growth factors may dictate the anabolic and chondrogenic activities of PRP.

Khatab et al. used a mouse model of OA to investigate the effects of PRPr on pain, cartilage damage, and synovial inflammation [24]. OA was induced by intra-articular injection of collagenase, and three consecutive intra-articular injections of PRPr or saline (control) were given to the affected knee. Weight distribution over the left and right hindlimbs was evaluated using an incapacitance tester as an indicator of pain. The positive effect of PRPr injection in reducing pain was demonstrated by the increased distribution of weight on the affected joint compared to the saline group. Moreover, at day 21, the PRPr group showed thinner synovial membrane and less cartilage damage than saline control. Histologically, the saline-injected joints showed significantly increased cartilage degeneration at day 28 compared to PRPr-injected groups. The data also showed a significant reduction in joint inflammation as evident from an increase in CD206+ and CD163+ anti-inflammatory macrophage cells suggesting the modulation of macrophage subtype by PRPr. The fewer number of pro-inflammatory macrophages in PRPr-injected knee points lowers prostaglandin E2 (PGE2) production in the knee. Since PGE2 is a lipid mediator of inflammatory pain, the study suggested that the significant reduction in pain can be attributed to the inhibition of PGE2 by tissue-resident macrophages. A related study has shown that PRPr contains high levels of interleukin 1 receptor antagonist (IL-1Ra) which has the ability to inhibit acute inflammation caused by IL-1 and promote macrophage polarization towards an anti-inflammatory (M2) phenotype [25].

The anti-inflammatory properties of PRP have also been partially attributed to its ability to reduce canonical NFkB signaling. Using cultured chondrocytes, Bendinelli et al. demonstrated the ability of $CaCl_2$-activated PRP to reduce the expression of NFkB transcriptional targets, cyclooxygenase 2 (Cox 2), and chemokine receptor CXCR4 [26]. The activated platelets showed enhanced release of HGF, IL-4, TNF-α, and TGF-β1. The study concluded that these growth factors play a key role in the anti-inflammatory effects of PRP. The suggested mechanism includes the reduction of NFkB expression and subsequently its transcriptional target Cox-2 and CXCR4 expression in chondrocytes by HGF and IL-4. HGF has shown to enhance NFkB inhibitor α (IKBα) expression which leads to the retention of NFkB-p65 subunit retention in the cytosol, thereby impairing its translocation to the nucleus, which is necessary for NFkB trans-activating activity and gene expression. In addition, HGF is also known to reduce PGE2 production by cells. Also, TGF-β1 has the potential to counteract TNF-α effect by preventing CXCR4 expression in monocytes and impeding monocyte chemotaxis towards its ligand.

Using an osteoarthritic mimic in vitro cell culture model, van Buul et al. demonstrated the ability of PRP to influence gene expression of matrix-forming and matrix-degrading proteins by reversing the IL-1β-induced inflammatory response in chondrocytes [27]. IL-1β-activated NFkB is known to turn off anabolic pathways such as type II collagen and aggrecan synthesis, and addition of PRP rescued the synthesis of these molecules by chondrocytes. However, no significant differences in the expression of matrix metalloproteinase-13 (MMP-13) gene and nitric oxide (NO) production by IL-1β treated chondrocytes in the presence of PRP. The study concluded that PRP has the ability to inhibit multiple inflammatory IL-1β-mediated effects in OA chondrocytes including inhibition of NFkB activation.

In addition to chondrocytes, PRP has been suggested to modify the synovial fluid as well as synoviocyte functions. Sundman et al. studied the effects of PRP on the expression of anabolic and catabolic genes as well as on the secretion of nociceptive and inflammatory mediators from OA cartilage and synoviocytes using an ex vivo co-culture model [28]. The study showed that PRP can enhance cellular metabolism (decrease catabolism) and at the same time decrease inflammatory and nociceptive markers such as TNF-α. Unlike the results of the previous study using chondrocytes [28], this study showed the ability of PRP to significantly decrease the expression of MMP-13 in synoviocytes. MMP-13 is known to play a key role in cartilage matrix degradation during the development and progression of OA. However, a study by Browning et al. showed the presence of multiple anabolic and catabolic mediators in PRP and that synoviocytes treated with PRP can respond with substantial MMP secretion [29]. These studies underline the fact that identifying and optimizing growth factors in PRP via standardizing its preparation methods are needed to improve function for specific applications. Interestingly, the study showed that PRP treatment led to an increased expression of HAS-2 in synoviocytes indicating its ability to increase HA production [28]. Anitua et al. corroborated the favorable effects of PRP by comparing the efficacy of platelet-rich and platelet-poor plasma (PPP) in enhancing the functions of synovial cells isolated from osteoarthritic patients. The study showed that PRP significantly enhanced HA secretion by synovial cells via HGF compared to PPP. The increased production of HA is significant as it could restore HA levels in the synovial fluid leading to joint lubrication and cartilage protection [30]. Lee et al. reported that PRP can significantly increase the expression of cannabinoid receptors (CB1 and CB2) in articular chondrocytes. Even though the exact role of these receptors in chondrocytes has not been fully elucidated, this may also act as a potential mechanism by which PRP modulate pain in osteoarthritis as well as impart chondroprotection.

In summary, these studies indicate that the favorable effects of PRP in reducing OA pain can be attributed to its ability to enhance joint metabolism and chondroprotective activity and decrease markers of inflammation (presumably via inhibiting NFkB) and nociception (by inhibiting tissue nociceptors) [31].

Efficacy of Combining PRP with Microfracture

Microfracture technique is commonly used to enhance repair of damaged cartilage by stimulating the healing process via drilling holes into the subchondral bone. Huh et al. evaluated the efficacy of combining microfracture with PRP for the treatment of chondral defect using a rabbit model [32]. At 12 weeks post-surgery, the PRP group showed almost complete defect coverage with neo-cartilage, and at the same time, the control group showed incomplete and irregular fibrous tissue formation. Significant differences in histological scoring were also observed demonstrating the beneficial effect of combining PRP with microfracture procedure. Milano et al. investigated the efficacy of PRP in microfracture procedure using a sheep full-thickness chondral lesion model [33]. The study confirmed using histological and macroscopic appearance score, the positive effect of PRP in improving cartilage repair and restoration. The study also found that the procedure is more effective when PRP was used as a gel compared to liquid intra-articular injection. However, the study showed that neither the control nor PRP group led to the formation of hyaline cartilage. Subsequently, another study was performed to evaluate the effect of repeated PRP injection in treating chondral lesion in the sheep model [34]. Two groups were included, wherein group 1 received 5 weekly injections of autologous conditioned plasma after microfracture and group 2 received only microfracture. The study showed significant improvements in group 1 based on macroscopic, histologic, and biomechanical analysis at 3, 6, and 12 months post-treatment. Group 1 showed significant improvement in tissue repair from 3 to 6 months and remained stable over time, whereas group 2 showed significant histologic and mechanical deterioration between 6 and 12 months. The study showed the feasibility of achieving more durable cartilage reparative response in the presence of PRP. Kruger et al. studied the effect of PRP in enhancing the migration and chondrogenic differentiation of subchondral progenitor cells to understand the efficacy of PRP therapy following microfracture [35]. Progenitor cells derived from human subchondral cortico-spongious bone were treated with PRP and analyzed for their migration and differentiation potential. The study showed that 0.1–100% PRP significantly stimulated progenitor cell migration compared to controls. Moreover, the progenitor cells stimulated by PRP showed increased cartilage matrix formation and enhanced chondrogenic gene markers. The study showed the potential of PRP to enhance the migration and differentiation of progenitor cells upon microfracture. These data point to the potential advantages of using PRP to promote cartilage tissue regeneration and delay joint degeneration.

Intra-Articular PRP Injection

The limitations of intra-articular corticosteroid and HA injections have led to the investigation of PRP as a potential alternative to address OA pain. Sampson et al. performed a pilot study to understand the efficacy of PRP using 14 patients with primary and secondary knee OA [36]. The patients received three intra-articular PRP injections at 4 weeks interval into the suprapatellar bursa of the affected knee using musculoskeletal ultrasound. The study reported lack of adverse events upon intra-articular PRP injections and significant and almost linear improvements in pain and symptoms. VAS scores showed significant improvement in pain upon knee movement and at rest, and the study reported favorable outcome at 12 months post-treatment indicating the potential of mid-term pain relief with intra-articular PRP injection. Another study evaluated the efficacy of intra-articular PRP injections at 6 and 12 months post-treatment in 100 patients (115 knees) with varying degree of articular degeneration (Kellgren 0-IV) [37]. Citrated blood was used, and an average of 6.8 million platelets was given to the lesion site at every injection. The first injection was given within 2 h of PRP preparation, and next two injections were administered every 21 days. Similar to the previous study, no major adverse events related to injections were observed. Also, statistically significant improvement was observed for all clinical scores (VAS, IKDC) from baseline to the end of the therapy and at 6–12 months follow-up. The study also indicated that younger patients showed much better effect compared to older patients. Similarly, the higher the extent of joint degeneration, the lower the effect. Subsequently, Filardo et al. performed a 2-year follow-up of the same population with 90 patients to understand the long-term efficacy of intra-articular PRP injection in reducing pain and functional outcome [38]. Compared to 12 months, the evaluated parameters worsened at 24 months follow-up. However, the evaluated parameters still showed significant improvement from the baseline showing the clinical advantages of the therapy to treat OA pain. In short, these earlier studies showed the potential of intra-articular PRP injection to reduce OA pain and warranted further studies including randomized clinical trials (RCTs) to clearly demonstrate its efficacy.

In 2015, Laudy et al. performed a comprehensive, systematic literature search of randomized or non-randomized controlled trials to determine the effect of intra-articular PRP (or similar products) injections on decreasing pain and improving functions, global assessment, and changes in joint images [39]. Ten clinical trials were included, and authors concluded that single or double PRP injections are more effective in reducing knee pain (limited level of evidence due to high risk of bias) compared to placebo control at 6 months post injection. Similarly, PRP showed a statistically significant effect in reducing pain compared to HA treatment (moderate level of evidence due to generally high risk of bias). The study showed the need for more large randomized clinical trials with low risk of bias to conclusively determine the efficacy of intra-articular PRP injections.

In 2017, Shen et al. performed a systematic review and meta-analysis of randomized controlled trials to determine the temporal effect of intra-articular PRP injection on pain and physical function and compared its efficacy to other intra-articular injections such as saline placebo, HA, ozone, and corticosteroids [40]. The meta-analysis comprised of 14 RCTs that included 1423 participants. Among these, four studies were considered as moderate risk of bias and ten as high risk of bias. The authors concluded that compared to the control groups, intra-articular PRP injections significantly reduced WOMAC pain scores at 3, 6, and 12 months follow-up. The WOMAC physical function subscores also showed significant improvement in the PRP groups compared to other groups at 3, 6, and 12 months post injection. Moreover, PRP injection did not significantly increase the risk of post injection adverse events. The conclusion of this study was supported by a Phase I and IIa clinical trial performed in Japanese patients with mild to moderate knee osteoarthritis [41]. Ten patients were enrolled in the study, and PRP without white blood cells (using a single-spin centrifuge from citrated blood) were injected intra-articularly. Three injections were given at 1 week intervals. The study showed that at 6 months, 80% of the patients showed 50% or more reduction in the VAS score indicating the efficacy of PRP treatment in reducing OA pain. Similarly, Wu et al. investigated the effect of PRP injection in reducing pain and improving muscle strength compared to placebo in bilateral knee OA [42]. Twenty patients (40 knees) were enrolled and received a single intra-articular injection of PRP (leukocyte- and platelet-rich plasma) or saline. The WOMAC was evaluated at baseline and at 2 weeks, 1, 3, and 6 months post injection. The muscle strength was evaluated using a calibrated iso-kinetic testing machine. Significant reduction in the WOMAC pain and total scores were observed in the PRP group compared to saline group at 1, 3, and 6 months post injection. In addition, the PRP group showed greater knee strength at a longer fol-low-up period; however, no significant differences in muscle strength were observed between the groups. In 2019, Chen et al. performed a summary of meta-analysis comparing intra-articular PRP injection to HA and placebo injections [43]. Four meta-analyses were included with level 1 evidence. Each individual meta-analysis included had a Quality of Reporting of Meta-analysis (QUOROM) score range from 14 to 17 points and Oxman-Guyatt (Oxman-Guyatt quality appraisal tool) score range from 4 to 6 points. Heterogeneity of the included studies are leukocyte-rich PRP (LR-PRP); leukocyte-poor PRP (LP-PRP); single or double spinning; activa-tion or not; PRP injection dose, times, and interval; as well as standardized and non-standardized patient outcome measures. Three meta-analyses showed increased benefit of PRP in OA pain relief and functional improvement compared to HA or placebo treatments, and one study showed no differences between the groups. This study once again confirmed the need for more rigorous RCTs focusing on specific questions such as PRP preparation method, activation method, concentration of platelets, composition of PRP, and frequency of PRP injections to clearly under-stand the clinical efficacy of intra-articular PRP injection to develop the optimal therapy. Also, more studies with mid-long-term follow-up are also needed to access the curative effect of PRP injection and its potential for long-term pain relief. Similar to Shen et al.'s study, no significant differences in adverse events between groups were observed. The study concluded that for short-term follow-up (≤ 1 year), intra-articular PRP injection is more effective in treating pain compared to HA or placebo. In 2020, Hohmann et al. performed a systematic review and meta-analysis to compare the efficacy of intra-articular injection of PRP to HA in reducing pain and improving functional outcomes at 6 and 12 months [44]. Level 1 and 2 studies were included with a minimum of 6 months follow-up. Twelve studies with 1248 cases (636 PRP and 612 HA) of symptomatic treatment of OA pain were included. Pain was assessed in the studies by VAS and WOMAC pain scores. Pooled estimate for pain demonstrated significant improvement in PRP group compared to HA group at 6 months and 12 months. Even though significant improvements in pain scores were observed, the results did not show significant improvements in func-tional outcomes. In 2020, Chen et al. performed a meta-analysis to compare the efficacy of PRP compared to HA intra-articular injections [45]. The meta-analysis comprised of 14 RCTs involving 1350 patients (714 in the PRP and 636 in the HA groups). The results showed the PRP group to be superior to the HA group in terms of mid- and long-term VAS (24 and 54 weeks) and long-term WOMAC pain score further confirming the efficacy of PRP compared to HA in reducing OA-associated pain. Yet another systematic review and meta-analysis of RCTs compared the effi-cacy and safety of PRP and HA injections [46]. This included 18 level 1 studies with 1608 patients (811 patients in PRP and 797 patients in HA groups) with a mean follow-up of 11.1 months. The results showed that the mean improvement of

WOMAC total score was significantly higher in the PRP group compared to HA. Similarly, 6 out of 11 studies reported VAS pain score indicating that PRP can significantly reduce pain compared to HA. Wu et al. recently performed a meta-analysis of RCTs with consistent treatment cycle and frequency of injection to compare the efficacy of intra-articular PRP to HA treatment [47]. Ten studies were included. The study showed the significant efficacy of PRP compared to HA in relieving pain and self-reported functional improvements based on VAS and WOMAC scores. Filardo et al. performed a meta-analysis of RCTs to evaluate the efficacy of intra-articular PRP injections to placebo and other intra-articular treatments [48]. The study included 34 RCTs published till January 2020. The study concluded that the effect of PRP goes beyond its mere placebo effect and showed statistically significant improvement in pain scores. Interestingly, it has been found that the beneficial effects of PRP increase with time and showed significant effects at 6 and 12 months after injection. This suggests the potential to consider PRP treatment as one that may have longer lasting effects rather than greater improvements. As discussed before, the moderate quality and the significant heterogeneity of the included studies highlight the need for further studies to investigate the appropriate injection dose, frequency, as well as interval time of PRP injection to develop the optimal therapeutic strategy for treating OA.

By now, these clinical trials showed the potential of intra-articular PRP in reducing pain in patients with early or moderate OA. Some of the recent studies therefore investigated the potential of intra-articular PRP injection for treating severe knee osteoarthritis by changing the mode of application of PRP. Sanchez et al. described a new technique of PRP infiltration for treating severe knee osteoarthritis [49]. Four patients with grade III or IV knee tibiofemoral osteoarthritis based on Ahlbäck scale were included in the study. Here, intraosseous infiltration of PRP into the subchondral bone is suggested along with intra-articular injection to favorably affect the deeper layers of the cartilage and subchondral bone. In a follow-up study, 14 patients between 40 and 77 years with severe knee osteoarthritis were treated with one intra-articular injection of PRP and two intraosseous injections of PRP. The osteoarthritis outcome score (KOOS) showed significant pain reduction from baseline at 24 weeks post-treatment. Statistically significant improvement was also observed in secondary outcomes such as VAS score ($p < 0.001$) and Lequesne Index ($p = 0.008$) [50]. However, the study has limitations due to the small sample size and the lack of an intra-articular injection alone control to demonstrate the significant improvement via the combination approach.

Currently there is a growing interest in characterizing the PRP composition of the various commercially available PRP preparations to identify the optimal composition for treating OA. One area of interest is in comparing the efficacy of leukocyte-rich (LR) and leukocyte-poor (LP) PRP in OA treatment. An in vitro study investigated the effect of LR-PRP and LP-PRP on synoviocytes and showed that LR-PRP caused significant cell death and led to a more pro-inflammatory cell phenotype [51]. Leukocytes in PRP have the potential to act as a source of pro-inflammatory stimuli and can also increase the expression of proteases such as elastases and MMPs. Previous studies have shown that MMP1 and MMP3 are produced by synoviocytes when cultured with LR-PRP. Using a rabbit OA model, Yin et al. demonstrated that intra-articular injections of LP-PRP can lead to reduction of PGE2 thereby reducing inflammation and pain associated with OA [52]. These studies show the differential effects of these preparations and warrant further optimization while developing intra-articular therapeutics. Subsequently, Filardo et al. performed a study involving 144 symptomatic subjects and showed that both LR-PRP and LP-PRP groups statistically improved functional scores at 2, 6, and 12 months with better results in younger subjects with lower degree of cartilage degeneration. However, it was noted that LR-PRP injection led to higher pain and swelling reaction compared to LP-PRP implying the need for larger RCTs to compare the efficacy of these two preparations for OA treatment [53].

Since growth factors in PRP may play a role in improving PRP-related outcome, Raeissadat et al. investigated the efficacy of PRP-derived growth factors (PRGF) in reducing pain in patients with symptomatic knee osteoarthritis compared to HA injection using a single-blinded randomized controlled trial. A total of 102 candidates (~58 years) were given 2 intra-articular PRGF injections 3 weeks apart or 3 weekly injections of HA in this 12 months study. At 12 months post injection, both PRGF and HA showed significant reduced WOMAC pain score compared to the baseline. The VAS pain score showed significantly reduced pain in the PRGF compared to the HA group at 12 months. Moreover, the study showed significant patient satisfaction in the PRGF group compared to the HA group implying the potential benefits of PRGF compared to HA. Limitations of this study, however, include lack of a placebo group and being an unblinded study [54]. Vaquerizo et al. conducted a multicenter randomized controlled clinical trial to investigate the efficacy of three injections of PRGF-Endoret compared to one single intra-articular injection of Duronale HA for reducing pain in patients with symptomatic knee OA. Using 96 patients (~63 years), the study showed that the PRGF-treated groups showed statistically significant improvement in pain score at both 24- and 48-week post injection compared to HA [55].

Studies so far indicate the potential beneficial effects of PRP, particularly in reducing pain in patients with mild to moderate osteoarthritis. However, the literature presents contradictory results particularly regarding the advantages of PRP in improving functional outcomes in OA patients. The American Academy of Orthopaedic Surgeons Clinical Practice Guidelines consortium gave an inconclusive recommendation and suggested that they could not recommend for or against PRP in the treatment of symptomatic knee OA [56]. Most of the clinical studies are limited by the differences in the RCT protocols and insufficient representation of some of the outcome indicators. Another major issues of these studies are the lack of standardization of study protocols particularly regarding the preparation methods of PRP, needle gauge for blood harvest and injection, concentration of platelets, activation status of platelets, leukocyte concentration, use of fresh or frozen PRP, anticoagulant use, local anesthetic use, palpation vs image guidance, injection volume, injection frequency, pre-injection and post injection protocol, type and severity of disease being treated, and patient-specific factors [57]. There is also a need for better designed clinical studies to confirm the effectiveness of intra-articular PRP, in order to rule out a placebo effect. These variables could play a critical role in determining the biological activity of PRP and hence can adversely affect our ability to achieve consistent analgesic effect.

Intra-Articular Injection of PRP and Hyaluronic Acid

Viscosupplementation by the intra-articular injection of HA is extensively used to manage osteoarthritic pain. HA is used for intra-articular injection as it is a major component of synovial fluid and acts as a lubricant and shock-absorbing agent through various biological and biomechanical mechanisms. However, studies have shown that the effectiveness of HA in reducing OA pain is short term and may not last more than a few months [18].

Intra-articular injection of PRP and hyaluronic acid are currently considered as two promising approaches to treat OA pain, even though their mechanisms of actions are quite different. PRP has shown to exhibit multiplicity of biological functions such as anti-inflammatory and chondroprotective properties due to the potent mixture of growth factors present in it, as well as exhibit endogenous analgesic effects to alleviate inflammation-related pain. The HA on the other hand has shown to increase the viscosity and elasticity of joint fluid and reduces pain via lubrication. Studies have shown positive therapeutic effect of intra-articular HA injection with initial efficacy at 4 weeks and peak effectiveness at 8 weeks which last for about 6 months [58], whereas clinical data with PRP injections show mid- to long-term effects with significant reduction of pain at 6 and 12 months post injection. In light of the evidences showing the potential advantages of both approaches, several studies investigated the efficacy of combining PRP with HA for intra-articular injection. The hypothesis here is that the protection of joints provided by HA combined with the anti-inflammatory, analgesic, and anabolic effects of PRP could lead to synergistic effects.

To understand the synergistic effect, a study investigated the mobility of synovial fibroblasts in HA and PRP combined with HA solution (HA + PRP) and showed significantly improved cell migration in the HA + PRP group, indicating that addition of PRP can increase the biological activity of HA [59]. A related study showed increased chondrocyte proliferation in HA + PRP solution compared to PRP alone demonstrating the synergistic effect of HA and PRP [60]. Russo et al. studied the rheological and biological properties of four different HA solutions with varying concentrations when combined with PRP. The study showed that PRP addition is not detrimental to the viscosupplementation effect of HA at high concentrations. Moreover, the presence of PRP promoted chondrocyte proliferation and glycosaminoglycan production confirming the potential for synergistic biological effect [61].

Chen et al. studied the synergistic effect of HA and PRP using an in vitro human chondrocyte culture model and an in vivo ACLT-OA model [62]. The in vitro study showed the ability of HA + PRP treatment to reduce pro-inflammatory cytokine expression and chondrocyte proliferation and enhance chondrocyte phenotype. The chondrocytes were isolated from five OA patients. The molecular mechanism of enhanced chondrogenesis by HA + PRP has been attributed to the cooperative activation of CD44 (by HA) and TGF-βIIR (by PRP), downstream mediators Smad2/3 and Erk1/2, and the chondrogenic transcription factor Sox9. The in vivo model showed the ability of PRP + HA to rescue meniscus tear and cartilage breakdown and reduce OA-related inflammation. The study concluded the ability of PRP + HA to synergistically promote cartilage regeneration and inhibit OA inflammation. Lana et al. performed a randomized controlled trail comparing HA, PRP, and sequential injection of HA and PRP for treating mild and moderate knee OA [63]. The study included 105 patients, and each patient received 3 intra-articular injections with 2-week intervals. Significant reduction in VAS pain score was observed in PRP group compared to HA group at 3, 6, and 12 months post injection. Combining HA and PRP (PRP + HA) showed significant decrease in pain compared to HA alone 1 year post-treatment and significantly increased physical function at 1 and 3 months when compared to PRP alone. The study concluded that combining PRP with HA could potentially lead to better functional outcomes. This was supported by a cross-sectional study with retrospective review of 64 patients which included 56 knees injected with HA and PRP and 45 knees with HA only [64]. The study showed significant improvement in pain scores in PRP + HA group compared to the PRP group. In 2015, Abate et al. studied the efficacy of combining intra-articular PRP injection with hyaluronic acid in patients with mild to moderate knee osteoarthritis. Forty patients were enrolled into two groups with one group receiving weekly injections of 2 mL of hyaluronic acid and 2 mL PRP and the other group receiving weekly only PRP (4–5 ml) injections for 3 weeks. The outcomes were evaluated at 1, 3, and 6 months using VAS, KOOS, and weekly NSAID consumption. Even though intra-group comparison showed significant improvements, infra-group comparison failed to demonstrate significant differences. The study showed that combination therapy has the same efficacy as PRP alone when administered at higher volume [65]. A randomized control trial using 105 patients with mild and moderate osteoarthritis on the other hand showed that combining hyaluronic acid with PRP resulted in significant decreases in pain ($P = 0.0001$) and functional limitations when compared to hyaluronic acid alone at 1 year follow-up [66]. This was corroborated by a recent study using 360 patients with four groups of double-blind treatment with PRP (2–14 mL); double-blind treatment with hyaluronic acid (0.1–0.3 mg); and combination therapy of PRP and HA and placebo group. Using WOMAC pain score, the study concluded that PRP treatment is significantly more effective than HA in reducing pain. Moreover, the combination treatment of PRP and hyaluronic acid (PRP + HA) improved pain and total WOMAC score compared to PRP and hyaluronic acid treatment alone. Zhao et al. performed a systemic review of meta-analysis of 5 RCTs and 2 cohort studies with a total of 941 patients to understand the efficacy of intra-articular injection of PRP combined with HA compared to PRP or HA alone to treat OA [67]. The study showed that the outcome of HA + PRP and PRP alone in relieving knee pain is similar at 1 and 3 months post injection. However, at 6 months post injection, VAS score and Lequesne index showed significant improvement in HA + PRP group compared to PRP alone group. The results demonstrated that HA combined with PRP may have promising clinical effects and from longer follow-up periods are required to determine its efficacy for long-term treatment. Renevier et al. evaluated the efficacy of "Cellular matrix™ PRP-HA" for the management of tibiofemoral knee osteoarthritis in patients who did not respond adequately to intra-articular HA injection [68]. The procedure involved preparing PRP in the presence of HA unlike the previous studies wherein sequential injection of PRP and HA was done. The study included 77 patients with grade II or III knee osteoarthritis and a pain at walking score between 3 and 8 on a numeric rating scale. The patients were given three intra-articular injections at D0, D60, and D180. The WOMAC score showed significant reduction in pain between the baseline and at D270. The pain reduction was observed irrespective of the body mass index (BMI) and the grade (II and III) of the OA. The treatment has shown to give long-lasting benefits for 50% of the patients and allowed avoiding surgery for almost 80% at 4 years follow-up.

These studies indicate the further benefits of combining intra-articular injection of HA with PRP to achieve long-term OA pain relief. Even though these initial studies show the beneficial effects, large RCTs with low risk of bias are needed to confirm the efficacy of the approach.

Conclusions

Despite the substantial differences in the preparation method, frequency, and number of injections and composition of PRP, positive treatment effects in terms of mid- to long-term pain reduction observed in preclinical and clinical studies highlight the potential advantage of intra-articular PRP therapy for treating OA pain. In addition to PRP alone, combination of HA with PRP (HA + PRP) also raises significant interest due to their potential synergistic effects. These promising clinical results warrant further fundamental studies to understand the mechanism of action of intra-articular PRP injection and optimizing PRP preparation methods for treating OA with different levels of joint degeneration.

References

1. Andia I. Platelet-rich plasma biology. In: Alves R, Grimalt R, editors. Clinical indications and treatment protocols with platelet-rich plasma in dermatology. Ediciones Mayo: Barcelona; 2016. p. 3–15.
2. Xu XR, Zhang D, Oswald BE, et al. Platelets are versatile cells: new discoveries in hemostasis, thrombosis, immune responses, tumor metastasis and beyond. Crit Rev Clin Lab Sci. 2016;53(6):409–30. https://doi.org/10.1080/10408363.2016.1200008.
3. Weyrich AS, Schwertz H, Kraiss LW, Zimmerman GA. Protein synthesis by platelets: historical and new perspectives. J Thromb Haemost. 2009;7:241–6.
4. Jonnalagadda D, Izu LT, Whiteheart SW. Platelet secretion is kinetically heterogeneous in an agonist-responsive manner. Blood. 2012;120:5209–16.
5. Surgenor DM, Wallace EL, Hao SHS, et al. Collection and transfusion of blood in the United States, 1982–1988. N Engl J Med. 1990;322:1646.
6. Dohan Ehrenfest DM, Rasmusson L, Albrektsson T. Classification of platelet concentrates: from pure platelet-rich plasma (P-PRP) to leucocyte- and platelet-rich fibrin (L-PRF). Trends Biotechnol. 2009;27:158–67.

7. Magalon J, Chateau AL, Bertrand B, Louis ML, Silvestre A, Giraudo L, et al. DEPA classification: a proposal for standardising PRP use and a retrospective application of available devices. BMJ Open Sport Exerc Med. 2016;2:e000060.
8. Fitzpatrick J, Bulsara MK, McCrory PR, Richardson MD, Zheng MH. Analysis of platelet-rich plasma extraction: variations in platelet and blood components between 4 common commercial kits. Orthop J Sports Med. 2017;5(1):2325967116675272.
9. Hsu WK, Mishra A, Rodeo SR, Fu F, Terry MA, Randelli P, et al. Platelet-rich plasma in orthopaedic applications: evidence-based recommendations for treatment. J Am Acad Orthop Surg. 2013;21:739–48.
10. Wallace IJ, Worthington S, Felson DT, Jurmain RD, Wren KT, Kaijanen H, Woods RJ, Lieberman DE. Knee osteoarthritis has doubled its prevalence since the mid-20th century. Proc Natl Acad Sci. 2017;114:9332–6.
11. Gabay O. Osteoarthritis: new perspectives. J Spine. 2012;1:e101.
12. Kon E, Filardo G, Drobnic M, Madry H, Jelic M, van Dijk N, Della Villa S. Non-surgical management of early knee osteoarthritis. Knee Surg Sports Traumatol Arthrosc. 2012;20(3):436–49.
13. Katz JN, Earp BE, Gomoll AH. Surgical management of osteoarthritis. Arthritis Care Res (Hoboken). 2010;62(9):1220–8.
14. Scarpignato C, Lanas A, Blandizzi C, et al. Safe prescribing of non-steroidal anti-inflammatory drugs in patients with osteoarthritis-an expert consensus addressing benefits as well as gastrointestinal and cardiovascular risks. BMC Med. 2015;13:55.
15. McAlindon TE, Bannuru RR, Sullivan MC, et al. OARSI guidelines for the non-surgical management of knee osteoarthritis. Osteoarthritis Cartilage. 2014;22:363–88.
16. Kompel AJ, Roemer FW, Murakami AM, Diaz LE, Crema MD, Guermazi A. Intra-articular corticosteroid injections in the hip and knee: perhaps not as safe as we thought? Radiology. 2019;293:656–63.
17. Bowman S, Awad ME, Hamrick MW, Hunter M, Fulzele S. Recent advances in hyaluronic acid based therapy for osteoarthritis. Clin Transl Med. 2018;7:6.
18. Trigkilidas D, Anand A. The effectiveness of hyaluronic acid intra-articular injections in managing osteoarthritic knee pain. Ann R Coll Surg Engl. 2013;95:545–51.
19. Smith PA. Intra-articular autologous conditioned plasma injections provide safe and efficacious treatment for knee osteoarthritis. Am J Sports Med. 2016;44:884–91.
20. Durant TJS, Dwyer CR, McCarthy MBR, Cote MP, Bradley JP, Mazzocca AD. Protective nature of platelet rich plasma against chondrocyte death when combined with corticosteroids or local anesthetics. Am J Sports Med. 2017;45:218–25.
21. Spreafico A, Chellini F, Frediani B, Bernardini G, Niccolini S, Serchi T, Collodel G, Paffetti A, Fossombroni V, Galeazzi M, Marcolongo R, Santucci A. Biochemical investigation of the effects of human platelet releasates on human articular chondrocytes. J Cell Biochem. 2009;108:1153–65.
22. Zaky SH, Ottonello A, Strada P, Cancedda R, Mastrogiacomo M. Platelet lysate favours in vitro expansion of human bone marrow stromal cells for bone and cartilage engineering. J Tissue Eng Regen Med. 2008;2:472–81.
23. Drengk A, Zapf A, Sturmer EK, Sturmer KM, Frosch K. Influence of platelet-rich plasma on chondrogenic differentiation and proliferation of chondrocytes and mesenchymal stem cells. Cells Tissues Organs. 2009;189:317–26.
24. Khatab S, van Buul GM, Kops N, Bastiaansen-Jenniskens YM, Bos K, Verhaar JA, van Osch GJ. Intra-articular injections of platelet-rich plasma releasate reduce pain and synovial inflammation in a mouse model of osteoarthritis. Am J Sports Med. 2018;46:977–86.
25. Luz-Crawford P, Djouad F, Toupet K, Bony C, Franquesa M, Hoogduijn M, Jorgensen C, Noel D. Mesenchymal stem cell derived interleukin 1 receptor antagonist promotes macrophage polarization and inhibits b cell differentiation. Stem Cells. 2016;34(2):483–92.
26. Bendinelli P, Matteucci E, Dogliotti G, Corsi MM, Banfi G, Maoni P, Desiderio MA. Molecular basis of anti-inflammatory action of platelet-rich plasma on human chondrocytes: mechanisms of NF-kappaB inhibition via HGF. J Cell Physiol. 2010;225:757–66.

27. Van Buul GM, Koevoet WLM, Kops N, Bos PK, Verhaar JAN, Weinans H, Bernsen MR, Osch GJVM. Platelet-rich plasma releasate inhibits inflammatory processes in osteoarthritic chondrocytes. Am J Sports Med. 2011;39:2362–70.
28. Sundman EA, Cole BJ, Karas V, Valle CD, Tetreault MW, Mohammed HO, Fortier LA. The anti-inflammatory and matrix restorative mechanisms of platelet-rich plasma in osteoarthritis. Am J Sports Med. 2014;42:35–41.
29. Browning SR, Weiser AM, Woolf N, Golish R, SanGiovanni TP, Scuderi GJ, Carballo C, Hanna LS. Platelet-rich plasma increases matrix metalloproteinases in cultures of human synovial fibroblasts. J Bone Joint Surg Am. 2012;94:e1721–7.
30. Anitua E, Sanchez M, Nurdem AT, Zalduendo MM, Fuente M, Azofra J, Andia I. Platelet-released growth factors enhance the secretion of hyaluronic acid and induce hepatocyte growth factor production by synovial fibroblasts from arthritic patients. Rheumatology (Oxford). 2007;46:1769–72.
31. Andia I, Maffulli N. Platelet-rich plasma for managing pain and inflammation in osteoarthrisis. Nat Rev Rheumatol. 2013;9:721–30.
32. Huh SW, Shetty AA, Kim SJ, Kim YJ, Choi NY, Jun YJ, Park IJ. The effect of platelet rich plasma combined with microfracture for the treatment of condral defect in a rabbit knee. J Tissue Eng Regen Med. 2014;11:178–85.
33. Milano G, Sanna Passino E, Deriu L, Careddu G, Manunta L, Manunta A, Saccomanno MF, Fabbriciani C. The effect of platelet rich plasma combined with microfractures on the treatment of chondral defects: an experimental study in a sheep model. Osteoarthr Cartil. 2010;18:971–80.
34. Milano G, Deriu L, Sanna Pssino E, Masala G, Manunta A, Postacchini R, Saccomanno MF, Fabbriciani C. Repeated platelet concentrate injections enhance reparative response of microfractures in the treatment of chondral defects of the knee: an experimental study in an animal model. Arthroscopy. 2012;28:688–701.
35. Krüger JP, Hondke S, Endres M, Pruss A, Siclari A, Kaps C. Human platelet-rich plasma stimulates migration and chondrogenic differentiation of human subchondral progenitor cells. J Orthop Res. 2012;30:845–52.
36. Sampson S, Reed M, Silvers H, Meng M, Mandelbaum B. Injection of platelet-rich plasma in patients with primary and secondary knee osteoarthritis: a pilot study. Am J Phys Med Rehabil. 2010;89(12):961–94.
37. Kon E, Buda R, Filardo G, Di Martino A, Timoncini A, Cenacchi A, Fornasari PM, Giannini S, Marcacci M. Platelet-rich plasma: intra-articular knee injections produced favorable results on degenerative cartilage lesions. Knee Surg Sports Traumatol Arthrosc. 2010;18:472–9.
38. Filardo G, Kon E, Buda R, et al. Platelet-rich plasma intra-articular injections for the treatment of degenerative cartilage lesions and osteoarthritis. Knee Surg Sports Traumatol Arthrosc. 2011;19(4):528–35.
39. Laudy ABN, Bakker EWP, Rekers M, Moen MH. Efficacy of platelet-rich plasma injections in osteoarthritis of the knee: a systematic review and meta-analysis. Br J Sports Med. 2015;49:657–72.
40. Shen L, Yuan T, Chen S, Xie X, Zhang C. The temporal effect of platelet rich plasma on pain and physical function in the treatment of knee oateoarthritis: systematic review and meta-analysis of randomized controlled trials. J Orthop Surg Res. 2017;12:16.
41. Taniguchi Y, Yoshioka T, Kanamori A, Aoto K, Sugaya H, Yamazaki M. Intra-articular platelet-rich plasma injections for treating knee pain associated with osteoarthritis of the knee in the Japanese population: a phase I and IIa clinical trial. Nagoya J Med Sci. 2018;80:39–51.
42. Wu Y, Hsu K, Li T, Chang C, Chen L. Effects of platelet rich plasma on pain and muscle strength in patients with knee osteoarthritis. Am J Phys Med Rehabil. 2018;97:248–54.
43. Chen P, Huang L, Ma Y, Zhang D, Zhang X, Zhou J, Ruan A, Wang Q. Intra-articular platelet – rich plasma injection for knee osteoarthritis: a summary of meta analyses. J Orthop Surg Res. 2019;14:385.
44. Hohmann E, Tetsworth K, Glatt V. Is platelet-rich plasma effective for the treatment of knee osteoarthritis? A systematic review and meta-analysis of level 1 and 2 randomized controlled trials. Eur J Orthop Surg Traumatol. 2020;30:955–67.

45. Chen Z, Wang C, You D, Zhao S, Zhu Z, Xu M. Platelet-rich plasma versus hyaluronic acid in the treatment of knee osteoarthritis. A meta-analysis. Medicine. 2020;99(11):e19388.

46. Belk JW, Kraeutler MJ, Houck DA, Goodrich JA, Dragoo JL, McCarty EC. Platelet-Rich Plasma versus hyaluronic acid for knee osteoarthritis: a systematic review and meta-analysis of randomized controlled trails. Am J Sports Med. 2021;49(1):249–60.

47. Wu Q, Luo X, Xiong Y, Liu G, Wang J, Chen X, Mi B. Platelet rich plasma versus hyaluronic acid in knee osteoarthritis: a meta-analysis with the consistent ratio of injection. J Orthop Surg. 2020;28:1–9.

48. Filardo G, Previtali D, Napoli F, Candrian C, Zaffagnini S, Grassi A. PRP injections for the treatment of knee osteoarthritis: a meta-analysis of randomized controlled trials. Cartilage. 2021;13(1_suppl):364S–75S.

49. Sanchez M, Fiz N, Guadilla J, et al. Intraosseous infiltration of platelet-rich plasma for severe knee osteoarthritis. Arthrosc Tech. 2014;3(6):e713–7.

50. Sanchez M, Delgado D, Sanchez P, et al. Combination of intra-articular and intraosseous injections of platelet rich plasma for severe knee osteoarthritis: a pilot study. Biomed Res Int. 2016;2016:4868613.

51. Assirelli E, Filardo G, Mariani E, Kon E, Roffi A, Vaccaro F, Marcacci M, Facchini A, Pulsatelli L. Effect of two different preparations of platelet-rich plasma on synoviocytes. Knee Surg Sports Traumatol Arthrosc. 2015;23:2690–703.

52. Yin WJ, Xu HT, Sheng JG, et al. Advantages of pure platelet-rich plasma compared with leukocyte- and platelet-rich plasma in treating rabbit knee osteoarthritis. Med Sci Monit. 2016;22:1280–90.

53. Filardo G, Kon E, MTP R, Vaccaro F, Guitaldi R, Martino AD, Cenacchi A, Fornasari PM, Marcacci M. Platelet-rich plasma intra-articular injections for cartilage degeneration and osteoarthritis: single- versus double-spinning approach. Knee Surg Sports Traumatol Arthrosc. 2012;20:2082–91.

54. Raeissadat SA, Ahangar AG, Rayegani SM, Sajjadi MM, Ebrahimpour A, Yavari P. Platelet rich plasma derived growth factor vs hyaluronic acid injection in the individuals with knee osteoarthritis: a one year randomized clinical trial. J Pain Res. 2020;13:1699–711.

55. Vaquerizo V, Plasencia A, Arribas I, Seijas R, Padilla S, Orive G, Anitua E. Comparison of inra-articular injections of plasma rich in growth factors (PRGF-Endoret) versus Durolane hyaluronic acid in the treatment of patients with symptomatic osteoarthritis: a randomized controlled trial. Arthroscopy. 2013;29:1635–43.

56. https://www.guidelinecentral.com/summaries/american-academy-of-orthopaedic-surgeons-clinical-practice-guideline-on-treatment-of-osteoarthritis-of-the-knee-2nd-edition/#section-420.

57. Pourcho AM, Smoth J, Wisniewski SJ, Sellon JL. Intraarticular platelet rich plasma injection in the treatment of knee osteoarthritis: review and recommendations. Am J Phys Med Rehabil. 2014;93:S108–21.

58. Bannuru RR, Natov NS, Dasi UR, Schmid CH, McAlindon TE. Therapeutic trajectory following intra- articular hyaluronic acid injection in knee osteoarthritis--meta-analysis. Osteoarthr Cartil. 2011;19(6):611–9.

59. Anitua E, Sanchez M, De la Fuente M, Zalduendo MM, Orive G. Plasma rich in growth factors (PRGF-Endoret) stimulates tendon and synovial fibroblasts migration and improves the biological properties of hyaluronic acid. Knee Surg Sports Traumatol Arthrosc. 2012;20(9):1657–65.

60. Marmotti A, Bruzzone M, Bonasia DE, Castoldi F, Rossi R, Piras L, Maiello A, Realmuto C, Peretti GM. One-step osteochondral repair with cartilage fragments in a composite scaffold. Knee Surg Sports Traumatol Arthrosc. 2012;20(12):2590–601.

61. Russo F, Este M, Vadala G, Cattani C, Papalia R, Alini M, Denaro V. Platelet rich plasma and hyaluronic acid blend for the treatment of osteoarthritis: rheological and biological evaluation. PLOS One. 2016; https://doi.org/10.1371/journal.pone.0157048.

62. Chen WH, Lo WC, Hsu WC, et al. Synergistic anabolic actions of hyaluronic plasma and platelet-rich plasma on cartilage regeneration in osteoarthritis therapy. Biomaterials. 2015;35:9599–607.

63. Lana JF, Weglein A, Sampson S, Vicente EF, Huber SC, Souza CV, Amback MA, Vincent H, Urban-Paffaro A, Onodera CMK, Annichino-Bizzacchi JM, Santana MHA, Belandero ED. Randomized controlled trial comparing hyaluronic acid, platelet-rich plasma and the combination of both in the treatment of mild and moderate osteoarthritis of the knee. JSRM. 2012;12(2):69.
64. Saturveithan C, Premganesh G, Fakhrizzaki S, Mahathir M, Karuna K, Rauf K, William H, Akmal H, Sivapathasundaram N, Jaspreet K. Intra-articular hyaluronic acid (HA) and platelet rich plasma (PRP) injection versus hyaluronic acid (HA) injection in grade III and IV knee osteoarthritis (OA) patients: a retrospective study on functional outcome. Malays Orthop J. 2016;10:35–40.
65. Abate M, Verna S, Schiavone C, et al. Efficacy and safety profile of a compound composed of platelet-rich plasma and hyaluronic acid in the treatment for knee osteoarthritis (preliminary results). Eur J Orthop Surg Traumatol. 2015;25:1321.
66. JFSD L, Weglein A, Sampson S, et al. Randomized controlled trial comparing hyaluronic acid, platelet-rich plasma and the combination of both in the treatment of mild and moderate osteoarthritis of the knee. J Stem Cell Regen Med. 2016;12:P69.
67. Zhao J, Huang H, Liang G, Zeng L, Yang W, Liu J. Effects and safety of the combination of platelet rich plasma and hyaluronic acid in the treatment of knee osteoarthritis: a systematic review and meta-analysis. BMC Musculoskelet Disord. 2020;21:224.
68. Renevier JL, Marc JF, Adam P. "Cellular matrix TM PRP-HA": a new treatment option with platelet-rich plasma and hyaluronic acid for patients with osteoarthritis having had an unsatisfactory clinical response to hyaluronic acid alone: results of a pilot multicenter French study with long-term follow-up. Int J Clin Rheumatol. 2018;13(4):226–9.

PRP in Hair Restoration

6

Keyur Naik and Elie M. Ferneini

Introduction

Hair plays an important role in an individual's appearance and self-perception. Humans have attempted to remedy hair loss as far back as 1500 BCE. The Ancient Egyptians first described a topical treatment for first hair loss using snake and crocodile fats, iron, and lead in the Ebers Papyrus [1]. More recently, a mixture of cold Indian tea and fresh lemon juice was widely used in the colonies of the British Empire in the 1850s [2]. Modern forms of hair loss treatment, including topical and oral medication as well as surgical hair transplantation, became popularized in the mid-twentieth century. Minoxidil, a topical solution or foam first developed as an oral antihypertensive, and finasteride, an oral medication used to treat benign prostatic hyperplasia, are the current standards of care for patterned hair loss treatment. However, the efficacy of these treatments is variable between users and dependent on consistency of use. Hair transplantation offers a surgical solution to hair loss. The hair transplant technique was first described by Japanese dermatologist Dr. Shoji Okuda in 1939 and later popularized in the United States by a New York City physician Dr. Norman Orentreich [3]. Today, the hair loss treatment industry is a 3.5 billion dollar industry in the United States alone, and it is poised to reach 5.5 billion dollars within the next decade [4].

K. Naik (✉)
Department of Oral and Maxillofacial Surgery, New York University Langone Medical Center/Bellevue Hospital Center, New York, NY, USA
e-mail: Keyur.Naik@nyulangone.org

E. M. Ferneini
Beau Visage Med Spa and Greater Waterbury OMS, Cheshire, CT, USA

Department of Surgery, Frank H Netter MD School of Medicine, Quinnipiac University, North Haven, CT, USA

Division of Oral and Maxillofacial Surgery, University of Connecticut, Farmington, CT, USA

The majority of treatments for hair loss target patients suffering patterned hair loss. Male pattern hair loss (MPHL), also known as androgenetic alopecia (AGA), affects 85% of men over the course of their lifetime. Androgens, particularly testosterone and dihydrotestosterone (DHT), cause hair follicle shrinkage and hair thinning in androgen-sensitive areas of the scalp. As a result, areas with previously thick, terminal hair are reduced to thin, vellus hair. The areas that are first affected are the frontal hairline and the vertex. Female pattern hair loss (FPHL) affects approximately 40% of women throughout their lifetime [5]. Authors have recently moved away from classifying FPHL as a form of AGA as the role of androgens in female hair loss is debated. Females experience a different pattern of hair loss than their male counterparts. FPHL demonstrates diffuse thinning and loss of density while the frontal hairline is usually maintained [6]. Stages of hair loss in males are most commonly classified by the Hamilton-Norwood scale, and female hair loss is classified by the Ludwig system [7]. These classification systems are utilized to determine severity of hair loss and responses to treatment. While treatment for patterned hair loss represents the bulk of hair loss therapy, other forms of hair loss can also be treated. Alopecia areata (AA) is an autoimmune form of hair loss seen commonly in children which can affect any part of the body and is patchy in nature. AA is significantly less prevalent than patterned hair loss with a prevalence of 0.1–0.2% and a lifetime incidence of 1.7% [8]. Other forms of hair loss include telogen effluvium, tinea capitis, trichotillomania, and androgen-dependent female hair loss, which are relatively less common. Indications and types of treatment vary for these forms of alopecia and frequently differ from patterned hair loss. Along with our increasing understanding of the types and causes of hair loss, forms of hair loss treatment are constantly evolving.

Platelet-rich plasma was first used as a form of hair loss treatment in the mid-2000s out of a growing interest around cell-based and regenerative medicine [9]. The use of platelet-rich plasma is well known to oral and maxillofacial surgeons, periodontists, and orthopedic surgeons for its use in post-surgical wound healing due to its anti-inflammatory properties. PRP products contain a high platelet concentration and harness the regenerative cytokines released by activated platelets in order to stimulate cell proliferation and increase collagen formation. PRP treatment for alopecia has grown in popularity as it is minimally invasive and autologous in nature. In addition, recovery time after the procedure is minimal, and there are few side effects. In this chapter, we will explore the mechanism of action of PRP, the evaluation and selection of patients for PRP therapy, the preparation and administration of PRP, and the outcomes of treatment particularly as they relate to use in patterned hair loss.

Plasma-Rich Plasma Mechanism of Action

Understanding the role of PRP in hair restoration requires a brief review of the life cycle of the hair follicle. Hair follicles undergo cyclical growth. In broad terms, the phases of the hair follicle life cycle are (1) neogen, a phase of regeneration and increasing growth; (2) anagen, a phase of maximum growth and active fiber production; (3) catagen, a phase of hair follicle regression and slowing growth; and (4) telogen, a phase of dormancy [10]. A hair follicle will repeat this cycle numerous times throughout its life. In AGA, when androgen-sensitive areas of the scalp are exposed to testosterone and DHT, the life cycle of the hair follicle changes. Androgens prevent the hair follicle from growing by increasing the length of the telogen phase and causing shrinkage of the follicle without scarring [11]. Female pattern hair loss results in similar non-scarring follicle shrinkage though the role of androgens is not clear.

PRP is thought to lengthen the life cycle of hair follicles and prevent follicle miniaturization. However, the exact mechanism of PRP in hair growth is not fully understood. Activated platelets are known to release a range of growth factors including platelet-derived growth factor (PDGF), transforming growth factor-beta (TGF-b), fibroblast growth factor (FGF), epidermal growth factor (EGF), and vascular endothelium growth factor (VEGF) that promote growth of the follicle and block apoptotic signaling [12–14]. These factors are thought to directly influence dermal papillae cells in the scalp. Dermal papillae contain multipotent stem cells responsible for the formation of the hair follicle. Growth factors released by activated PRP increase the transcription of genes that cause proliferation and differentiation of dermal papillae stem cells [13, 15]. Certain growth factors released by PRP have specific roles in the hair follicle life cycle. EGF and hepatocyte growth factor (HGF) speed up the progression to the anagen phase. Insulin-like growth factor (IGF-1) is required for anagen maintenance [9]. Extending the anagen phase prolongs active growth of the hair follicle.

Platelet-rich plasma also improves hair follicle health indirectly by promoting angiogenesis and vascularization in the scalp [15]. VEGF is particularly responsible for neovascularization after injection of PRP into the scalp. This finding has been reproduced in both animal and human immunohistochemistry studies that have measured microvessel density and demonstrated a significant increase in density in areas of hair that have undergone PRP injections versus control groups. Robust blood supply to the follicle is essential for initiation of the anagen phase and the development of new follicles [16, 17]. Conversely, poor vascularization is noted on histologic specimens of scalps experiencing AGA [9, 18]. As such, PRP results in hair growth directly by increasing cell survival and proliferation of stem cells in the dermal papillae and indirectly by promoting angiogenesis and neovascularization.

Patient Evaluation and Selection

The patient evaluation begins with a comprehensive medical history and physical examination. Understanding the time course, severity, pattern, and associated symptoms surrounding the patient's hair loss is important for distinguishing the different forms of alopecia. Asking about hair thinning versus hair shedding, hair loss from the root versus broken hair, and patterned hair loss versus patchy hair loss are a few of the preliminary questions a provider can ask in order to determine the type of hair loss. A history of metabolic, infectious, endocrine, and autoimmune disorders as well as a drug, psychosocial, nutritional, and family history can provide additional information in order to separate different forms of hair loss [19]. A thorough medical history can also help identify patients in which PRP is contraindicated. Absolute contraindications are few; however, harvesting PRP requires phlebotomy and patients who are not eligible for phlebotomy due to platelet dyscrasias or local infection at the site cannot undergo PRP hair restoration therapy [20].

The clinical exam should include evaluation of the entire scalp as well as other hair bearing areas of the body. Particular attention should be given to the pattern and distribution of hair loss. Additionally, the quality and density of hair in multiple areas of the scalp including the frontal hairline, frontal scalp, parietal scalp, temporal scalp, and vertex should be evaluated. During the physical exam, magnification should be used to look for the presence of scar formation in order to distinguish scarring versus non-scarring forms of hair loss [19]. Physical maneuvers such as the hair pull test and the tug test can evaluate for shedding and fragility. It is important to evaluate the quality and density of hair outside of the scalp. The patient can be asked about and examined for changes in eyebrow, facial, or body hair. Presence of hair loss or thinning outside of the scalp may point to systemic disorders that should be further evaluated. The process of a detailed workup and the algorithms for diagnosing hair loss patterns are beyond the scope of this chapter. However, this description should be used as an introduction to the evaluation and examination of patients presenting for PRP hair loss treatment.

Prior to treatment, patients should be informed of possible complications and side effects. Due to the minimally invasive and autologous nature of the treatment, the side effect profile is limited. Some patients note postoperative headaches and erythema and edema around the injection sites [21].

Preparation and Administration

Currently, there are no standardized protocols for the preparation of platelet-rich plasma for hair restoration. Broadly, there are four types of preparations for isolating platelet products: pure platelet-rich plasma (P-PRP), leukocyte- and platelet-rich plasma (L-PRP), pure platelet-rich fibrin (P-PRF), and leukocyte- and platelet-rich fibrin (L-PRF) [22]. The principle differences lie in the concentrates of the platelets, fibrin, and leukocytes in these products. P-PRP and L-PRP are the two preparations most used in the hair restoration treatment. Certain protocols use activators in order to increase growth factor production. Thrombin and calcium chloride are examples of such activators. However, maintaining platelet stability and reducing activation prior to injection is critical for maximum benefit from the released platelet factors. Use of anticoagulants such as heparin is frequently added to preparations in order to balance activation with platelet stability [23]. Additionally, greater concentrations of platelets do not necessarily yield better outcomes, and beyond certain concentrations, angiogenesis may be impaired. One study reported an effective therapeutic concentration of 1.5×10^6 platelets/uL, which requires an enrichment of approximately two to six times baseline [24]. In order to standardize protocols and optimize treatment outcomes, the effects of different preparations, use of activators, and platelet density on hair restoration should be further explored.

The technique for administration of PRP varies between clinicians. The preparation can be injected using intradermal or subcutaneous injection. The volumes administered and the surface area covered are typically titrated to severity of hair loss. Number and frequency of treatments are also determined by the practitioner. Wide ranges of treatment number (between 1 and 6 treatments) and frequencies (between 1 week and 3-month intervals) are noted in the literature. Improvement in hair density and quality is typically appreciable over the course of 3 months after completion of treatment.

Patient Outcomes

In large part, platelet-rich plasma outcome studies have demonstrated positive results though these studies are typically limited to patients undergoing PRP treatment for patterned hair loss. A number of studies have demonstrated a statistically significant improvement in hair density and diameter in patient suffering from AGA using noninvasive or minimally invasive forms of recipient site sampling and analysis after treatment. Studies varied in the number of treatments of PRP; however, the majority that showed positive outcomes used multiple treatments over a 3-month period. Gkini et al. also demonstrated that patients with Norwood grade II–III had greater improvements in hair thickness than Norwood grades V–VII [25]. The presence of vellus hair was associated with more robust results (compared to areas with no hair). As such, patients with milder forms of hair loss and more vellus hair had better responses to treatment, and they are likely better candidates for treatment.

Conversely, two studies found no difference in hair density and thickness outcomes after PRP treatment. Mapar et al. performed a split-scalp study with 19 male patients suffering from AGA and demonstrated no statistically significant improvement in the areas of scalp that received PRP as opposed to placebo saline. However, it should be noted that Mapar's treatment protocol only called for two administrations 1 month apart, which is considerably fewer than most studies that reported positive results [26]. Puig et al. performed a large study on the use of PRP in females with patterned hair loss with no significant improvement in hair count or hair mass index [27]. These studies highlight the need for standardized trials in order to optimize patient selection and treatment protocols.

While the efficacy of PRP in hair restoration continues to be debated, patient satisfaction with the procedure remains high. Multiple studies have evaluated patient-centered measures, including self-perceived hair quality and density, rate of hair fall, and overall satisfaction with the procedure. Studies that involved patient self-assessment questionnaires have shown a perceived complete cessation in hair fall of in between 75% and 90% of patients and an increase in hair growth, hair quality, and appearance in approximately 65% [28, 29]. Two studies that assessed overall patient satisfaction both reported scores of greater than 7 out of 10 [25, 30].

Today, quality of life (QOL) is one of the driving parameters used to compare and recommend medical procedures. QOL is particularly important when evaluating cosmetic procedures. The high patient satisfaction with PRP hair restoration is an encouraging sign of PRP's utility in this space.

Platelet-rich plasma has also been compared to and used with other forms of hair loss treatment. Verma and colleagues designed a trial to compare outcome of PRP treatment to minoxidil with a 6-month follow-up among 30 patients. PRP outperformed minoxidil in quantitative measures such as the hair tug test and also patient-centered self-evaluations and an overall patient satisfaction survey [31]. Minoxidil is a topical hair loss treatment that is recommended for use twice daily to the scalp. Patient compliance with this regimen is highly variable. In addition, minoxidil is associated with a number of side effects, the most common being scalp irritation, growth of hair in undesirable areas, and, rarely, headache and lightheadedness. Finasteride, an oral medication used in the treatment of AGA, also has a number of side effects including loss of libido and impotence. PRP limited side effect profile, and the need for few treatments improves compliance against those seeking non-surgical hair loss treatments. PRP has also been studied when used in conjunction with surgical hair transplantation, specifically follicular unit extraction (FUE). Two studies, a split-scalp study and a randomized control trial, both found greater follicular graft density in the treatment group that received PRP with FUE compared to FUE only, demonstrating the utility of PRP as an adjuvant therapy in hair transplantation [32, 33].

Conclusions

Platelet-rich plasma is a new biotechnology that is now used as a form of injectable treatment for hair loss. The therapy is growing in popularity as a treatment for hair loss due to its autologous nature, low side effect profile, and relative low cost compared to surgical treatment such as hair transplant. Unlike minoxidil and finasteride, lifelong application is not required. PRP is also not dependent on the patient's current amount of donor hair unlike hair transplantation and is unlikely to result in scar formation. Patient selection, preparation of PRP, and technical details of the procedure are not well studied and are not standardized with guidelines.

Most of our understanding of the mechanism of action of PRP comes from its extensive use in wound healing. While the mechanism of action of PRP is fairly well understood, its application as a hair loss therapy is still debated. Most current studies focus on male patterned hair loss and very few on female pattern hair loss. PRP's use in other forms of hair loss has not been studied. Early studies demonstrate largely favorable results using split-scalp study designs or control groups, though studies demonstrating no significant difference have also been reported. All studies were based on a small number of patients, and the treatment protocol and forms of evaluation varied between studies. Despite these limitations, the preliminary studies surrounding the use of PRP in hair restoration are largely favorable.

Ultimately, PRP is quickly growing in popularity as a hair loss treatment. Patients are increasing searching for "lunchtime" cosmetic procedures with limited downtime. PRP promises to be that form of treatment for hair loss. However, larger studies are required in order to standardize and further study the efficacy of PRP hair restoration.

References

1. Cunha F. I. The Ebers Papyrus. Am J Surg. 1949;77(1):134–6. https://doi.org/10.1016/0002-9610(49)90394-3.
2. Yesudian P. Can beverages grow hair on bald heads? Int J Trichol. 2012;4(1):1. https://doi.org/10.4103/0974-7753.96078.
3. Shiell R. A review of modern surgical hair restoration techniques. J Cutan Aesthet Surg. 2008;1(1):12. https://doi.org/10.4103/0974-2077.41150.
4. Santos Z, et al. Drug discovery for alopecia: gone today, hair tomorrow. Expert Opin Drug Discovery. 2015;10(3):269–92.
5. Monselise A, et al. What ages hair. Int J Womens Dermatol. 2015;1(4):161–6. https://doi.org/10.1016/j.ijwd.2015.07.004.
6. Ghanaat M. Types of hair loss and treatment options, including the novel low-level light therapy and its proposed mechanism. South Med J. 2010;103(9):917–21. https://doi.org/10.1097/smj.0b013e3181ebcf71.
7. Mysore V, Gupta M. Classifications of patterned hair loss. Hair Transplant. 2016:45. https://doi.org/10.5005/jp/books/12881_9.
8. Pratt CH, et al. Alopecia areata. Nat Rev Dis Primers. 2017;3:17011. https://doi.org/10.1038/nrdp.2017.11.
9. Stevens J, Khetarpal S. Platelet-rich plasma for androgenetic alopecia: a review of the literature and proposed treatment protocol. Int J Women's Dermatol. 2018;5(1):46–51. https://doi.org/10.1016/j.ijwd.2018.08.004.
10. Bernard BA. Advances in understanding hair growth. F1000Res. 2016;5:F1000 Faculty Rev-147. https://doi.org/10.12688/f1000research.7520.1.
11. Li ZJ, Choi HI, Choi DK, Sohn KC, Im M, Seo YJ. Autologous platelet-rich plasma: a potential therapeutic tool for promoting hair growth. Dermatol Surg. 2012;38(7 Pt 1):1040–6.
12. Randall VA. Androgens and hair growth. Dermatol Ther. 2008;21:314–28. https://doi.org/10.1111/j.1529-8019.2008.00214.x.
13. Alves R, Grimalt R. A review of platelet-rich plasma: history, biology, mechanism of action, and classification. Skin Appendage Disord. 2018;4(1):18–24.
14. Uebel CO, da Silva JB, Cantarelli D, Martins P. The role of platelet plasma growth factors in male pattern baldness surgery. Plast Reconstr Surg. 2006;118(6):1458–66; discussion 1467.
15. Driskell RR, et al. Hair follicle dermal papilla cells at a glance. J Cell Sci. 2011;124(Pt 8):1179–82. https://doi.org/10.1242/jcs.082446.
16. Goldman BE, Fisher DM, Ringler SL. Transcutaneous PO2 of the scalp in male pattern baldness: a new piece to the puzzle. Plast Reconstr Surg. 1996;97(6):1109–16; discussion 1117.
17. Mecklenburg L, Tobin DJ, Müller-Röver S, Handjiski B, Wendt G, Peters EM. Active hair growth (anagen) is associated with angiogenesis. J Invest Dermatol. 2000;114(5):909–16.
18. Takikawa M, et al. Enhanced effect of platelet-rich plasma containing a new carrier on hair growth. Dermatol Surg. 2011;37(12):1721–9. https://doi.org/10.1111/j.1524-4725.2011.02123.x.
19. Mubki T, Rudnicka L, Olszewska M, Shapiro J. Evaluation and diagnosis of the hair loss patient: part I. History and clinical examination. J Am Acad Dermatol. 2014;71: 415.e1–15.

20. Garg S, Manchanda S. Platelet-rich plasma-an 'Elixir' for treatment of alopecia: personal experience on 117 patients with review of literature. Stem Cell Investig. 2017;4:64. https://doi.org/10.21037/sci.2017.06.07.
21. Cervantes J, Perper M, Wong LL, Eber AE, Villasante Fricke AC, Wikramanayake TC, Jimenez JJ. Effectiveness of platelet-rich plasma for androgenetic alopecia: a review of the literature. Skin Appendage Disord. 2018;4:1–11. https://doi.org/10.1159/000477671.
22. Ehrenfest DMD, et al. In search of a consensus terminology in the field of platelet concentrates for surgical use: platelet-rich plasma (PRP), platelet-rich fibrin (PRF), fibrin gel polymerization and leukocytes. Curr Pharm Biotechnol. 2012;13(7):1131–7. https://doi.org/10.2174/138920112800624328.
23. Badran KW, Sand JP. Platelet-rich plasma for hair loss: review of methods and results. Facial Plast Surg Clin North Am. 2018;26(4):469–85.
24. Giusti I, Rughetti A, D'Ascenzo S, Millimaggi D, Pavan A, Dell'Orso L. Identification of an optimal concentration of platelet gel for promoting angiogenesis in human endothelial cells. Transfusion. 2009;49(4):771–8.
25. Gkini MA, Kouskoukis AE, Tripsianis G, Rigopoulos D, Kouskoukis K. Study of platelet-rich plasma injections in the treatment of androgenetic alopecia through an one-year period. J Cutan Aesthet Surg. 2014;7(4):213–9.
26. Mapar MA, et al. Efficacy of platelet-rich plasma in the treatment of androgenetic (male-patterned) alopecia: a pilot randomized controlled trial. J Cosmet Laser Ther. 2016;18(8):452–5. https://doi.org/10.1080/14764172.2016.1225963.
27. Puig CJ, et al. Double-blind, placebo-controlled pilot study on the use of platelet-rich plasma in women with female androgenetic alopecia. Dermatol Surg. 2016;42(11):1243–7. https://doi.org/10.1097/dss.0000000000000883.
28. Anitua E, Pino A, Martinez N, Orive G, Berridi D. The effect of plasma rich in growth factors on pattern hair loss: a pilot study. Dermatol Surg. 2017;43:658–70.
29. Gupta S, Revathi TN, Sacchidanand S, Nataraj HV. A study of the efficacy of platelet-rich plasma in the treatment of androgenetic alopecia in males. Indian J Dermatol Venereol Leprol. 2017;83:412.
30. Khatu SS, More YE, Gokhale NR, Chavhan DC, Bendsure N. Platelet-rich plasma in androgenic alopecia: myth or an effective tool. J Cutan Aesthet Surg. 2014;7:107–10.
31. Verma K, et al. A study to compare the efficacy of platelet-rich plasma and minoxidil therapy for the treatment of androgenetic alopecia. Int J Trichology. 2019;11(2):68. https://doi.org/10.4103/ijt.ijt_64_18.
32. Uebel CO, et al. The role of platelet plasma growth factors in male pattern baldness surgery. Plast Reconstr Surg. 2006;118(6):1458–66. https://doi.org/10.1097/01.prs.0000239560.29172.33.
33. Garg S. Outcome of intra-operative injected platelet-rich plasma therapy during follicular unit extraction hair transplant: a prospective randomised study in forty patients. J Cutan Aesthet Surg. 2016;9(3):157. https://doi.org/10.4103/0974-2077.191657.

PRP in Facial Rejuvenation

7

Kevin C. Lee and Elie M. Ferneini

Introduction

The goal of facial rejuvenation is to reverse age-related skin changes. Over time, both intrinsic and extrinsic processes act to thin the epidermis and reduce the natural volume of collagen and subcutaneous fat. The net result is a thin, translucent skin that suffers from decreased strength, elasticity, and regenerative potential. Facial rejuvenation can be achieved through a wide variety of surgical and nonsurgical treatments. Some procedures, such as rhytidectomy and dermal fillers, directly alter or augment the aging tissues. Other procedures, such as microneedling and skin resurfacing, repurpose wound healing principles to combat natural aging.

Platelets play a critical role in facilitating the inflammatory and proliferative phases of wound healing. As a result, they are often harvested for their regenerative properties. Platelet-rich plasma (PRP) products are estimated to contain four to seven times the platelet concentration of normal human blood. There are four categories of PRP formulations that are divided on the basis leukocyte and fibrin content. PRP refers to products with a low fibrin content, whereas platelet-rich fibrin (PRF) refers to products with a high fibrin content. The differences in preparation are beyond the scope of this chapter and are covered elsewhere. PRP is extracted as a fluid; however, benefits of increased platelet density can be augmented through activation with calcium chloride or thrombin. Platelet activation causes

K. C. Lee (✉)
Division of Oral and Maxillofacial Surgery, New York-Presbyterian/Columbia University Irving Medical Center, New York, NY, USA
e-mail: kcl2136@cumc.columbia.edu

E. M. Ferneini
Beau Visage Med Spa and Greater Waterbury OMS, Cheshire, CT, USA

Department of Surgery, Frank H Netter MD School of Medicine, Quinnipiac University, North Haven, CT, USA

Division of Oral and Maxillofacial Surgery, University of Connecticut, Farmington, CT, USA

© The Author(s), under exclusive license to Springer Nature Switzerland AG 2022
E. M. Ferneini et al. (eds.), *Platelet Rich Plasma in Medicine*,
https://doi.org/10.1007/978-3-030-94269-4_7

degranulation and the release of multiple adhesion molecules and growth factors. The presence of extracellular adhesion molecules causes activated PRP to have a gelatinous consistency similar to that of PRF. Inactivated PRP is preferred for most facial rejuvenation procedures because the liquid phase is more versatile and easier to deliver subcutaneously. Regenerative cytokines such as platelet-derived growth factor (PDGF), transforming growth factor-beta (TGF-b), fibroblast growth factor (FGF), epidermal growth factor (EGF), and vascular endothelium growth factor (VEGF) are responsible for enhancing collagen synthesis, stimulating cellular proliferation, and increasing chemotaxis. Few studies have been able to quantify the concentration of these growth factors.

The added value of PRP to facial rejuvenation procedures is a matter of scientific debate. When used as an adjunct in facial rejuvenation, PRP is theorized to decrease postprocedural downtime, reduce the incidence of complications, and improve the final cosmetic outcome. There is truly limited evidence to support all of these purported benefits of PRP; however, this is not an admission of failure. Because the clinical application of PRP has only been recently described, much of the evidence derives from in vitro studies [1]. We anticipate that future trials will allow us to distill the fact from the fiction. The purpose of this chapter is to provide an overview of noninvasive facial rejuvenation procedures and to summarize the current evidence underlying the use of PRP.

PRP for Facial Injections

PRP or PRF alone can be injected intradermally or subdermally to serve as a natural, bioactive filler. The proposed mechanism of action is similar to that of other applications. Namely, PRP products are thought to activate dermal fibroblasts and increase collagen and hyaluronic acid deposition. Extracellular matrix remodeling with PRP has been demonstrated in vitro. As previously discussed, PRF is more viscous because of the cross-linked fibrin framework. Unlike hyaluronic fillers, PRF is not easily dissolvable with an antidote, and particular care should be taken to aspirate prior to each injection. The number of treatments and length of time between injections is a matter of individual philosophy. It has been recommended that three to five injections be performed 4–6 weeks apart to achieve a loading response followed by a tailored maintenance regimen [2]. When hyaluronic acid fillers are combined with PRP, the procedure is termed a "vampire facelift." Proponents believe that adding PRP will promote cellular differentiation and dermal growth to sustain volume even after the hyaluronic acid disappears.

Some of the best evidence regarding PRP injections comes from the split-face randomized controlled trial conducted by Alam et al. [3]. The investigators recruited patients with cheek rhytids of at least Glogau class II and injected these blinded patients intradermally with 3 mL of normal saline and PRP. Although masked dermatologists were unable to discern differences in pigmentation or texture between the control and PRP sides, patients self-reported significantly decreased wrinkling on the PRP side at 6-month follow-up. Likewise, Yuksel et al. also found that patient ratings were more favorable than dermatologist ratings [4]. In their study, 1.5 mL of PRP was applied to ten patients in the forehead, malar area, and jaw by a dermaroller and in the lateral canthal rhytids with a 27-gauge injector. This regimen was performed three times at 2-week intervals. Three months after the last injection, the dermatologists only identified significant improvements in skin firmness, whereas the patients all claimed significant improvements in general appearance, skin firmness, and wrinkles. The duration of any volume changes is unclear, and it is important to note that they may only become visibly apparent to the patient months after injection. Of note, both studies assessed raters for pigmentation changes, but no discernable changes were identified. To the best of our knowledge, there is no evidence outside of case reports to suggest that PRP alone causes pigmentation changes [5]. Some authors have proposed using PRP injections in the management of infraorbital dark circles [6]. Dark circles can be causes by either excess pigmentation or skin translucency. PRP was not found to affect melanin content, and therefore PRP may treat dark circles through dermal thickening [6].

We were only able to identify one study in the cosmetic literature evaluating the effects of combining soft tissue fillers and PRP. Good aesthetic results with minimal complications were reported among 75 patients treated with the combination of hyaluronic acid and PRP, suggesting that adding PRP may be safe and does not worsen the cosmetic effect of fillers [7]. Multiple studies in the orthopedic literature have investigated the use of hyaluronic acid and PRP for osteoarthritis, and they have likewise reported few complications with the admixture. Still, it is unclear how the cosmetic outcomes with PRP compare to filler alone. It is also unclear if PRP truly sustains the augmented filler volume.

Laser Skin Resurfacing

Ablative lasers disrupt the skin surface by vaporizing portions of the epidermis and the included pigment-producing cells. Fractionated lasers damage narrow columns of tissue called microthermal zones (MTZs) in an evenly spaced pattern across the treated surface. Adjacent healthy tissues repopulate these MTZs. Lasers also produce heat in the underlying dermis which causes collagen contracture and deposition. The resulting pigmentation reduction and dermal tightening are used to reverse photoaging. Erythema, edema, and crust generally last for 1 week. Laser therapy is also associated with a variety of longer lasting, untoward complications including both hyper- and hypopigmentation and deeper dermal scarring [8]. PRP can be combined with laser therapy as a preoperative injection, as a postoperative injection, or as a topical solution over the ablated skin. When applied topically, a 20- to 30-min contact time with a soaked gauze carrier or an activated PRP gel is typically used. It is unclear which, if any, delivery method is superior. A handful of studies have evaluated the value of adjuvant PRP in both reducing post-laser downtime/complications and improving the final cosmetic outcome.

The best evidence for adjuvant PRP comes from split-face placebo-controlled trials. Lee et al. performed the only split-face randomized trial for post-laser PRP injections [9]. Following resurfacing for acne scars, the authors injected 6 mL of saline in the control side and 6 mL of PRP in the experimental side. Erythema, edema, and crusting all resolved sooner on the PRP side by 1–2 days. After a second laser PRP session, the final cosmetic outcome at 4 months was rated by two blinded dermatologists. The PRP side was rated to have a marginally superior cosmetic improvement compared to the control side. Hui and colleagues similarly performed a randomized split-face trial but instead used PRP (~2.2 mL) in the experimental arm as both a pre-treatment injection and a post-laser gauze-soaked mask. On the control side, normal saline was used for both injection and topical coating. Their laser PRP regimen involved three sessions spread over a 3-month interval after which time the effects were rated by two blinded dermatologists. As with the prior study, erythema, edema, and crusting were reported to have resolved 1–2 days sooner on the PRP side. The final cosmetic outcome likewise showed that the PRP side had marginal, but statistically significant, improvements in wrinkle, texture, and elasticity scores. A few well-designed trials have evaluated the isolated effects of MTZ-facilitated PRP absorption. Kar and Raj performed a split-face study evaluating ablative lasers and topical PRP, and they followed the same three-session monthly regiment as prior studies [10]. Although topical PRP reduced the severity of post-laser symptoms, there were no significant improvements in the final cosmetic outcome. Shin et al. compared combined laser and topical PRP and showed that adjuvant PRP lengthened the dermal-epidermal junction and increased both the collagen content and the number of fibroblasts [11]. Although their patients were significantly more satisfied with PRP and although histologic changes were present, the blinded dermatologists reported no differences in objective improvement scores.

Although PRP injections themselves do not alter melanin content [6], it is unclear if PRP can help reduce the incidence of unanticipated hypo- or hyperpigmentation complications with lasers. Among 158 patients treated with lasers and PRP, no patients had hyperpigmentation or depigmentation [12]. Abdel Aal et al. performed a split-face laser and intradermal PRP study, and they found five cases of hyperpigmentation none of which occurred on the PRP sides. On a microscopic basis, Na et al. found that adding PRP to laser treatment significantly reduced the melanin index [13]. In contrast, Shin et al. reported no difference in melanocyte index with PRP [11]. No definitive evidence supports the ability of PRP to reducing pigmentation changes with laser therapy.

Microneedling

Microneedling causes focal damage to the papillary and reticular dermis in order to induce collagen formation. Fine dermal rollers or oscillating needles that penetrate up to 3 mm in depth can be used for this purpose. Topical PRP applied to an intact skin surface is not thought to have a clinical benefit because growth factors are unable to efficiently access the vascularized dermis. Along with ablative laser treatment, needling is one method for delivering PRP to the dermis. PRP can be delivered before or after needling as an intradermal injection or as a topical application.

When applied topically, the duration of skin contact varies, but like with lasers, a 20- to 30-min contact time is recommended. However, this is not a firm recommendation, and some have used activated PRP gel as a post-needling topical agent for up to an hour [14]. Creative techniques have also been described for depositing liquid PRP interprocedurally [15]. Combining microneedling with PRP is colloquially termed the "vampire facial." The downtime with microneedling is less than that of lasers because there is minimal epidermal trauma. Likewise, pigmentation and scarring are less of a concern. Therefore, in microneedling, adjuvant PRP is primarily used to improve the cosmetic outcome.

Like with all other procedures, the evidence for PRP in microneedling is based on a scattering of small sample studies. El-Domyati et al. looked at needling with adjuvant PRP for post-acne scarring [16]. They found that 5 min of post-treatment topical PRP improved the histology of dermal structures compared to dermaroller alone at 3 months following the last of six treatment sessions. Ibrahim et al. also performed a split-face study for post-acne scarring [17]. Topical activated PRP was applied to the experimental arm after microneedling. There were no significant differences in cosmetic improvement between the two study arms; however, the PRP group had approximately a 2-day decrease in the duration of post-needling edema and erythema.

Fat Grafting

Autologous fat grafts are used as natural filler materials to treat volume loss in the aging face. Fat grafts can be placed in the periorbital region, temporal fossa, malar eminence, lips, and nasolabial folds. Because fat has an unpredictable survival rate, there is a tendency to overgraft in anticipation of resorption. Overgrafting is not without complication, and fat necrosis, calcification, and even accidental overcorrection are possible sequelae. A variety of techniques, such as adjusting cannula diameter, have been proposed to optimize fat graft survival [18]. Along those lines, some providers have recently proposed using adjuvant PRP with the transplanted fat to promote vascular ingrowth and differentiation of mesenchymal precursor cells [19]. Certainly the ability of growth factors to promote the survival of non-vascularized grafts has been well-described in the maxillofacial literature with bone morphogenetic proteins [20]. Like with many other PRP applications, there is an abundance of in vitro evidence supporting the ability of PRP to increase fall cell survival [21, 22]. In rats, a 20% mixture of PRP was shown to increase capillary ingrowth and help maintain normal adipocyte morphology up to 120 days after transfer [23].

The combination of PRP with autologous fat is relatively straightforward; however, the amount and preparation of PRP are important considerations. As with all other applications, the therapeutic concentration of PRP is a matter of debate. Commonly used "dosages" of inactivated PRP are generally between 20% and 40% by volume. Because there is a concern that higher concentrations of PRP inhibit adipogenic differentiation, some have advocated for activated PRP admixtures as low as 5% [24]. Unlike other adjuvant applications, fat grafting does not necessarily require liquid phase PRP, and this flexibility can be used to a patient's advantage. Activated PRP has a gelatinous consistency which may be a preferable carrier medium when more extensive soft tissue reconstruction is required, such as when treating progressive hemifacial atrophy (Parry-Romberg syndrome) [25]. Furthermore, PRF appears to increase tissue retention better than PRP. Xiong et al. compared microscopic findings 12 weeks after transfer and found that the neovascular density of fat grafts was higher with PRF than PRP [26]. PRF has been shown to release more growth factors over an extended period of time. The authors of the study hypothesized that the three-dimensional fibrin framework of PRF supported a scaffold for cellular ingrowth. They also proposed that activating PRP may blunt the benefits of PRP because fat healing is a protracted process and activation decreases the effective life span of the PRP growth factors.

Conclusion

Multiple split-face randomized trials have been conducted to assess the efficacy of PRP products in facial rejuvenation; however, much of the evidence is based on limited sample sizes. Post-treatment PRP appears to reduce the duration of edema and erythema by 1–2 days. The cosmetic benefits appear to be real based on the consistency of results across multiple published studies. Specifically, for fat grafting, a gelatinous inactivated formulation of PRP may optimize graft survival. Alleged complications with PRP are minimal and were rarely reported. Standard facial rejuvenation procedures inherently carry greater treatment risks. Common to all PRP applications, there is wide variability in the preparation and content of products. This chapter presents general principles and available treatment options; however, the heterogeneity of protocols makes it impossible to provide precise evidence-based treatment recommendations.

References

1. Kim DH, Je YJ, Kim CD, Lee YH, Seo YJ, Lee JH, et al. Can platelet-rich plasma be used for skin rejuvenation? Evaluation of effects of platelet-rich plasma on human dermal fibroblast. Ann Dermatol. 2011;23(4):424–31.
2. Peng GL. Platelet-rich plasma for skin rejuvenation: facts, fiction, and pearls for practice. Facial Plast Surg Clin North Am. 2019;27(3):405–11.
3. Alam M, Hughart R, Champlain A, Geisler A, Paghdal K, Whiting D, et al. Effect of platelet-rich plasma injection for rejuvenation of photoaged facial skin: a randomized clinical trial. JAMA Dermatol. 2018;154(12):1447–52.
4. Yuksel EP, Sahin G, Aydin F, Senturk N, Turanli AY. Evaluation of effects of platelet-rich plasma on human facial skin. J Cosmet Laser Ther. 2014;16(5):206–8.
5. Uysal CA, Ertas NM. Platelet-rich plasma increases pigmentation. J Craniofac Surg. 2017;28(8):e793.
6. Mehryan P, Zartab H, Rajabi A, Pazhoohi N, Firooz A. Assessment of efficacy of platelet-rich plasma (PRP) on infraorbital dark circles and crow's feet wrinkles. J Cosmet Dermatol. 2014;13(1):72–8.

7. Lee H, Yoon K, Lee M. Full-face augmentation using Tissuefill mixed with platelet-rich plasma: "Q.O.Fill". J Cosmet Laser Ther. 2019;21(3):166–70.
8. Halepas S, Lee KC, Higham ZL, Ferneini EM. A 20-year analysis of adverse events and litigation with light-based skin resurfacing procedures. J Oral Maxillofac Surg. 2020;78(4):619–28.
9. Lee JW, Kim BJ, Kim MN, Mun SK. The efficacy of autologous platelet rich plasma combined with ablative carbon dioxide fractional resurfacing for acne scars: a simultaneous split-face trial. Dermatol Surg. 2011;37(7):931–8.
10. Kar BR, Raj C. Fractional CO2 laser vs fractional CO2 with topical platelet-rich plasma in the treatment of acne scars: a split-face comparison trial. J Cutan Aesthet Surg. 2017;10(3):136–44.
11. Shin MK, Lee JH, Lee SJ, Kim NI. Platelet-rich plasma combined with fractional laser therapy for skin rejuvenation. Dermatol Surg. 2012;38(4):623–30.
12. Cai J, Tian J, Chen K, Cheng LH, Xuan M, Cheng B. Erbium fractional laser irradiation combined with autologous platelet-rich plasma and platelet-poor plasma application for facial rejuvenation. J Cosmet Dermatol. 2020;19(8):1975–9.
13. Na JI, Choi JW, Choi HR, Jeong JB, Park KC, Youn SW, et al. Rapid healing and reduced erythema after ablative fractional carbon dioxide laser resurfacing combined with the application of autologous platelet-rich plasma. Dermatol Surg. 2011;37(4):463–8.
14. Asif M, Kanodia S, Singh K. Combined autologous platelet-rich plasma with microneedling verses microneedling with distilled water in the treatment of atrophic acne scars: a concurrent split-face study. J Cosmet Dermatol. 2016;15(4):434–43.
15. Pathania V, Oberoi B, Shankar P, Bhatt S. Single-handed vampire facial: combining microneedling with platelet-rich plasma for single-hand use. J Am Acad Dermatol. 2021;84(2):e77–8.
16. El-Domyati M, Abdel-Wahab H, Hossam A. Microneedling combined with platelet-rich plasma or trichloroacetic acid peeling for management of acne scarring: a split-face clinical and histologic comparison. J Cosmet Dermatol. 2018;17(1):73–83.
17. Ibrahim MK, Ibrahim SM, Salem AM. Skin microneedling plus platelet-rich plasma versus skin microneedling alone in the treatment of atrophic post acne scars: a split face comparative study. J Dermatolog Treat. 2018;29(3):281–6.
18. James IB, Bourne DA, DiBernardo G, Wang SS, Gusenoff JA, Marra K, et al. The architecture of fat grafting II: impact of cannula diameter. Plast Reconstr Surg. 2018;142(5):1219–25.
19. Modarressi A. Platlet rich plasma (PRP) improves fat grafting outcomes. World J Plast Surg. 2013;2(1):6–13.
20. Herford AS, Miller M, Signorino F. Maxillofacial defects and the use of growth factors. Oral Maxillofac Surg Clin North Am. 2017;29(1):75–88.
21. Rodriguez-Flores J, Palomar-Gallego MA, Enguita-Valls AB, Rodriguez-Peralto JL, Torres J. Influence of platelet-rich plasma on the histologic characteristics of the autologous fat graft to the upper lip of rabbits. Aesthet Plast Surg. 2011;35(4):480–6.
22. Pires Fraga MF, Nishio RT, Ishikawa RS, Perin LF, Helene A Jr, Malheiros CA. Increased survival of free fat grafts with platelet-rich plasma in rabbits. J Plast Reconstr Aesthet Surg. 2010;63(12):e818–22.
23. Nakamura S, Ishihara M, Takikawa M, Murakami K, Kishimoto S, Nakamura S, et al. Platelet-rich plasma (PRP) promotes survival of fat-grafts in rats. Ann Plast Surg. 2010;65(1):101–6.
24. Kakudo N, Minakata T, Mitsui T, Kushida S, Notodihardjo FZ, Kusumoto K. Proliferation-promoting effect of platelet-rich plasma on human adipose-derived stem cells and human dermal fibroblasts. Plast Reconstr Surg. 2008;122(5):1352–60.
25. Cervelli V, Gentile P. Use of cell fat mixed with platelet gel in progressive hemifacial atrophy. Aesthet Plast Surg. 2009;33(1):22–7.
26. Xiong S, Qiu L, Su Y, Zheng H, Yi C. Platelet-rich plasma and platelet-rich fibrin enhance the outcomes of fat grafting: a comparative study. Plast Reconstr Surg. 2019;143(6):1201e–12e.

PRP in Oral and Maxillofacial Surgery and Dental Implants

8

Steven Halepas, Xun Joy Chen, and Alia Koch

Introduction

Oral and Maxillofacial surgery (OMS) covers a wide variety of procedures, from dentoalveolar surgeries such as extractions and dental implants, to soft tissue reconstruction, to orthognathic jaw surgery. The American Association of Oral and Maxillofacial Surgeons tries to explain the scope to dental students by dividing into six categories, dentoalveolar surgery, infections, pathology, trauma, orthognathic surgery, cleft/craniofacial reconstruction, TMJ, and facial cosmetics. The field, thus, has long searched for methods to aid in post-surgical healing, remodeling, and regeneration.

When teeth are extracted, bone remodeling is essential to the recovery process. Third molar extraction is a common procedure in Oral and Maxillofacial surgery. Approximately ten million wisdom teeth are extracted from five million patients every year in the United States [1]. Similarly, dental implants are a part of daily practice and often the ideal treatment for replacing missing teeth. Brânemark et al. first described the use of the modern-day implant in North America in 1982 [2]. With technological advancements, such as cone beam computed tomography (CBCT) and intra-oral scanning, practitioners can plan the treatment of more advance cases and have improved clinical outcomes. More complex cases often require technically demanding hard and/or soft tissue augmentation. As the need for bone augmentation increases, the dental literature has become flooded with numerous techniques and materials. One, of recent excitement, is the use of platelet-rich plasma and platelet-rich fibrin. Marx first described the use of platelet-rich plasma

S. Halepas (✉) · A. Koch
Division of Oral and Maxillofacial Surgery, New York-Presbyterian/Columbia University Irving Medical Center, New York, NY, USA
e-mail: sh3808@cumc.columbia.edu

X. J. Chen
Will Surgical Arts, Rockville, MD, USA

© The Author(s), under exclusive license to Springer Nature Switzerland AG 2022
E. M. Ferneini et al. (eds.), *Platelet Rich Plasma in Medicine*,
https://doi.org/10.1007/978-3-030-94269-4_8

(PRP) and platelet-rich fibrin (PRF) in the dental field in 1998 where he reported positive healing of the alveolar bone with its use [3]. Platelet-rich plasma/fibrin has since gained substantial popularity in the dental and medical community.

PRP is a concentration of platelet and plasma proteins derived from whole blood that is placed in a centrifuge to remove the red blood cells. PRP production involves the use of anticoagulants. PRF, however, is made without the use of anticoagulants, and the blood is immediately placed into the centrifuge after phlebotomy. This allows the coagulation cascade to occur causing a PRF matrix to be formed in the test tube that traps cytokines and other growth factors (see Chap. 3, PRP Preparation). PRP is believed to work via the degranulation of the alpha granules in platelets which contain several growth factors [4]. PRP contains a variety of growth factors/cytokines such as transforming growth factor beta (TGF-beta), platelet-derived growth factor (PDGF), insulin-like growth factor (IGF), and epidermal growth factor (EGF). While both PRP and PRF contain many of the same growth factors, much research is being invested into the quantity. Given the matrix formed in PRF, it is believed that the growth factors are released slowly over time as compared to PRP, but some of the amount of these cytokines may be lost in the process. A study preformed observed the growth factors released from PRP and PRF over a period. The highest produced growth factors were PDGF-AA, PDGF-BB, TGFB1, VEGF, and PDGF-AB. PRP was demonstrated to provide the highest growth factor in the short term. Over a 10-day period, however, they noted that PRF released the highest amounts of total growth factor [5].

It is well known that PDGF/IGF-1, when added to bone defects and implants, promotes osteoblasts and osteointegration [6]. Although numerous studies have demonstrated statistical significance from a biological basis, none to our knowledge have formulated a concrete clinical difference [7, 8]. A systematic review was conducted to determine the efficacy of PRP for non-transfusion use in the fields of dentistry, orthopedics, and wound care. The authors retrieved a total of 1240 references and found statistically significant results regarding the use of PRP in treatment of intra-bony defects as well as bone augmentation for implant placement. However, the authors in this study noted a high risk of bias in the individual studies due to lack of randomization and blinding [9].

Platelet-rich products are being utilized and studied in each of the disciplines of OMS. This chapter will explore its use in dentoalveolar (specifically extractions, ridge augmentation, and dental implants), in reconstruction (cleft/craniofacial, bone and soft tissue defects), as well as in pathology (medication-related osteonecrosis of the jaw) and TMJ disorders. Its use in facial cosmetics will be explored in other chapters of this book (see Chap. 5, Hair Restoration, and Chap. 6, Facial Rejuvenation).

Dentoalveolar

Extraction Sockets

PRP and PRF have been hypothesized to aid in bone healing after extraction sockets. Since extraction of third molars is so common, they are useful models for investigating this theory. Mandibular third molar extractions specifically are generally more painful and have more postoperative complications than maxillary third molars. PRF has long thought to provide benefit in regard to reducing the incidents of alveolar osteitis, pain, trismus, and swelling following extraction of third molars. The reason for this is due to the enhanced healing by the growth factors. In 2017 a study was published in the Journal of Oral and Maxillofacial Surgery in which they conducted a systematic review and meta-analysis to determine the efficacy of platelet-rich fibrin after third molar extractions [10]. They concluded that limited evidence exists and there is a need for standardized randomized controlled trials to truly determine the efficacy of PRF. The meta-analysis only found six papers published in this topic with all of them from outside of the United States. There has been a huge push for the use of platelet-rich fibrin in all aspects of medicine with dentistry included.

A Cochrane review was published in 2020 that included 62 trials and 4643 participants to explore surgical techniques of mandibular third molar extractions. The authors found "lacing platelet-rich plasma (PRP) or platelet-rich fibrin (PRF) in sockets may reduce the incidence of alveolar osteitis (OR 0.39, 95% CI 0.22–0.67; 2 studies) [11]." Zhu et al. published a meta-analysis in 2021 that included 42 studies, stating that PRF significantly reduced the incidence of both alveolar osteitis and postoperative pain [12]. Malhotra et al. preformed a study exploring bone regeneration in extraction sockets of third molars and found that faster bone formation in the PRF sockets compared to the control [13]. The data is still imperfect in regard to clear benefit; it appears biologically plausible that platelet-rich products are beneficial in the wound healing process of third molar extractions.

Implant Osteointegration

A systematic review conducted in 2014 determined the 10-year survival rate of dental implants is approximately 94.6% [14]. As clinicians, it is hard to ask for better odds, yet as dentistry advances, patients are expecting a guarantee. Current research is looking at how to shorten the integration time and time to prosthesis. PRP and PRF can possibly serve a role in this purpose.

A split mouth randomized clinical trial was conducted involving PRF and implants in the posterior maxilla. Implant stability was determined using resonance frequency analysis at 2, 4, and 6 weeks after placement. The authors found statistically significant increased ISQ at the 6-week period with implants placed using PRF [15]. The study is limited in that it only includes 20 patients. While the statistically significant result does not necessarily correlate to meaning clinically significant, it does demonstrate that biologically, PRF is having some sort of measurable effect. Another study performed using 72 dental implants in 9 beagles attempted to analyze the bone remodeling using PRP and PRF. After 3-month follow-up, the authors concluded that there was no increase in primary or secondary implant stability, but they did see a biological improvement in the peri-implant bone volume and structural integration [16].

In a randomized, single-blinded, controlled clinical trial, involving placing 41 immediate implants, half received PRF at the peri-implant region and half did not; the authors found no increased implant stability when using radiofrequency analysis [17]. Although clinical effects have yet to be established, a biological effect is being consistently observed. In one in vitro study in which roughened titanium dental implants were treated with PRP, the authors found that the number of cells observed around the implant at day 5 was double that of the non-PRP coated [18]. In another study where titanium implants were placed in the femurs of rats using PRF, the authors found that the PRF caused an osseoinductive response as compared to the control [19].

Keratinized Soft Tissue

The utilization of PRF has been associated with better epithelialization and improved soft tissue healing. The growth factors that constitute PRP and PRF promote fibroblasts and other healing mechanisms. With this information it is logical that PRP and PRF cause a better soft tissue response after surgical procedure. Many studies have demonstrated both a faster remodeling and an increased thickness in keratinized mucosa when using PRP or PRF. A randomized, split-mouth design was conducted for eight patients who needed bilateral widening of keratinized mucosa around dental implants in the mandible. On one side of the mouth, a free gingival graft was placed, while on the other, a PRF membrane was placed. The mean amount of keratinized mucosa at the implant at the PRF-only site was 3.3 mm ± 0.9 and 3.8 mm ± 1.0 at the free gingival graft site [20]. The use of PRF membranes may provide an alternative to restoring the keratinized gingiva around implants. PRF membranes are a great alternative, being less invasive, do not require a donor site, and have less postoperative pain.

On a study involving 126 immediately placed dental implants and the use of PRP, the authors found a statistically significant soft tissue healing score as compared to the control at 3 and 7 days. The study does note no difference found at 5-year follow-up, however [21]. In the exploration of soft tissue healing, many studies also found decrease pain and swelling associated with PRP and PRF at the surgical procedure. While both of these tend to be difficult to measure and are relatively subjective, any benefit could be an added bonus for the patient. As experienced providers know, keratinized gingiva is very important at dental implant sites and difficult to get back if lost. Maintaining as much keratinized gingiva as possible is very valuable when placing dental implants.

Sinus Lifts

The maxillary bone tends to resorb in the apical and palatal direction. In the posterior maxilla, the lack of vertical bone height may prevent implant placement. When minimal vertical bone height exists between the crest of the edentulous ridge and the maxillary sinus, the provider may need to perform a sinus lift [22].

Internal sinus lifts can be performed when there is already at least 5 mm of bone between the ridge and the maxillary sinus. There must be adequate bone height below the sinus for the implant to be stable when it is inserted in the alveolar bone. This approach is executed by making the osteotomy just short of the sinus floor. Hand osteotomes can then be used to "up-fracture" the remaining cortical bone at the sinus floor, and graft material can be packed at the osteotomy site [23, 24]. Minimizing the extent of the elevation decreases the risk of creating a hole in the sinus floor and perforating the sinus membrane. Some proponents have argued just packing a PRF pellet at osteotomy site and pushing it to the base of the sinus membrane before putting in the implant [25]. The PRF pellet might potentially be a good alternative to packing bone particulate graft material or conjunction.

An external sinus lift is performed using a window that is created in the lateral sinus wall after reflecting an extensive mucoperiosteal flap often involving releasing incisions. After flap elevation, the sinus may be visible through the lateral wall showing a transparent/blue appearance. A rectangular- or oval-shaped window is created with a large round diamond bur or now more commonly with an ultrasonic piezotome using great care to prevent perforating the underlying sinus membrane [26]. Utilization of a Dentium Advanced Sinus Kit or other sinus membrane kits can also aid in forming the osteo-window without sinus communication. Bone graft material is placed into the new space created between the apical aspect of the edentulous ridge and the sinus. The provider should wait 5–6 months after sinus augmentation and then 3 months after implant placement prior to prosthetic treatment. If the sinus elevation is not extensive, and enough bone height already exists to achieve primary stability, the implant can be placed at the time of the sinus augmentation. Some have advocated the use of PRF membranes either along the sinus membrane or along the osteotomy window defect (Fig. 8.1).

A recent study looked at the use of Unilab Surgibone with and without PRP on bone healing of the sinus floor. They measured histologic and residency frequency analysis for implant stability. Bone biopsies were conducted at time of implant placement on average 7 months following the sinus augmentation. Out of the ten

Fig. 8.1 PRF membranes

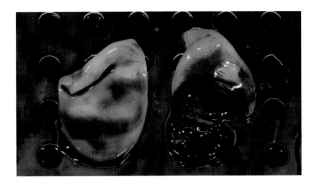

patients, they found no statistically significant effect in either measure [27]. A Cochrane systematic review in 2010 found no statistical efficacy for its use in dental implant sinus lifts but did note limited data existed [28].

Ridge Augmentation

Bone quality is an important factor in successful osteointegration. Lekohlm and Zarb classify bone quality into four categories depending on the ratio of compact bone and spongy bone as well as the subjective bone resistance when drilling [29] (Table 8.1).

Bone grafting procedures are utilized to augment the region generally when insufficient bone exists at a potential implant site. Sometimes bone grafting is used when there is sufficient bone to stabilize an implant but not enough to cover all the implants surfaces [30, 31]. Guided bone regeneration (GBR) generally consists of utilizing bone particulate substances. Bone augmentation can be done months prior to implant placement to facilitate future implant placement. The primary purpose of these procedures is to provide a scaffold and maintain volume for new bone formation.

Allografts are biomaterials generally composed of cadaver bone. An important difference from autogenous bone is the loss of endogenous cells and growth factors resulting from processing and sterilizing the allograft, thereby reducing or eliminating osteoinductive capacity. Consequently, the allograft will serve as an effective osteoconductive scaffold for new bone regeneration [32]. PRP is believed to act as a reservoir for stem cells and therefore provide that osteoinductive capacity that is found in autologous grafts (see Fig. 8.2). A xenograft is a graft that is obtained from

Table 8.1 Lekohlm and Zarb bone quality classification system

Type 1 bone	Very hard and dense and has a less prominent blood supply
Type 2 bone	Thick layer of compact one with a core of dense trabecular bone
Type 3 bone	A thin layer of compact bone surrounding a core of trabecular bone
Type 4 bone	The spongiest with a thin layer of cortical bone around a core of low density trabecular bone

Fig. 8.2 Platelet-rich plasma mixed with bone particulate

another species. The xenografts used in alveolar bone regeneration are osteoconductive bone graft substitute made of inorganic components but derived from non-human sources. The use of bone graft substitutes eliminates the need for harvesting of autogenous bone or harvesting a smaller volume of autogenous bone while providing structure and slow resorption characteristics that stabilize and "protect" the autogenous component of the bone graft. In addition, the use of bone graft substitutes eliminates donor site morbidity. To take best advantage of these characteristics, clinicians often mix and/or layer autogenous bone and biomaterial bone graft substitute components [33]. It is expected that implants can be placed into previously placed block or particulate bone grafts 5–6 months post graft placement; however, there is older data that shows implant stability may be more predictable when placed 6–9 months following bone grafting [34–37]. When particulate grafting occurs simultaneous with implant placement, the implant will often begin prosthetic treatment approximately 3–4 months after implant placement [38].

It has been demonstrated that PRF gradually releases autologous growth factors and is more effective for the proliferation of differentiation of osteoblasts than PRP in vivo [39]. PRF can act as a resorbable membrane for guided bone regeneration in that it prevents the emigration of non-desirable cells into the bone defect as well as provide a source of healing by promoting cytokines [40]. Therefore, the use of PRF can be used as a membrane itself or in conjunction with other membranes. An additional use of PRF with bone particulate is the fabrication of "sticky bone" which is PRF and allograft. The fibrin acts as a glue to hold the particulate together and can be molded into the desired morphology (see Fig. 8.3). The protocol for preparation of sticky bone is described in Chap. 8, PRP Cases.

Reconstruction

Clefts

In a similar fashion to ridge augmentation, PRP is being investigated for benefit in alveolar cleft bone grafting. Sakio et al. reported on a study 29 patients with unilateral alveolar clefts, with 6 in the control, and 23 in the PRP group. All patients underwent iliac cancellous bone grafts. The authors [41] found no significant difference in mean resorption of the bone on follow-up imaging, suggesting that PRP does not decrease bone resorption following the procedure. A similar study was preformed 3 years later by Chen and colleges. They aimed to analyze the newly formed bone volume 6 months after secondary alveoloplasty using iliac cancellous bone graft, with and without platelet-rich plasma in 40 patients, this time with 20 patients in each group. The authors found no statistical difference in bone formation on postoperative bone volume [42]. The current literature, therefore, does not show benefit with PRP in terms of bone volume. It would be helpful if other studies explored any soft tissue benefit when using PRP or if the alveolar cleft that was grafted with PRP has increased dental implant survival long term.

Fig. 8.3 Creation of sticky bone using PRF and bony particulate. Venipuncture is preformed to obtain 10 ccs of blood and placed in the centrifuge at 2500 rpms for 3.5 min. The top plasma/fibrin layer is removed with an empty syringe which is approximately 1 cc. Generally, you can expect to obtain 1 cc of autologous fibrin glue for each 10 cc test tube of venous blood. Mix the autologous fibrin glue immediately with bone particles on a metal dish. Mold the bone into the desired thickness with a periosteal elevator. Leave the coated bone undisturbed on the metal dish for 5–10 min. The bone particulate will form "sticky bone" (PRF + allograft)

Large Bone Defects

In addition to being used in small maxillary and mandibular ridge augmentations, PRP is being used in large bone defects in conjunction with other reconstruction techniques. Custom cribs of titanium mesh are being fabricated to help facilitate bone growth in maxillary and mandibular reconstruction. Marx et al. have described this in detail using both bone particular (CCFDAB) and bone harvested from anterior iliac crests along with rhBMP-2 to stimulate bone formation [43]. In addition to the rhBMP-2, Marx et al. incorporate PRP. Marx and colleagues define the "tissue engineering triangle" as a source of cells, a signal, and a matrix. The recombinant human bone morphogenic protein-2 (rhBMP-2) is a chemoattractant providing the stem cells, crushed cancellous freeze-dried allogenic bone (CCFDAB) provides the osteoconductive matrix, and the PRP provides the signal through its growth factors [43]. PRF membranes have also been used to lay over the titanium mesh to aid in soft tissue closure. PRF can also be used after the mesh is removed, laying over the newly formed bone.

Pathology

The growth factors of PRP have thought to be beneficial in cases of osteonecrosis of the jaws (ONJ). Osteoradionecrosis of the jaws (ORNJ) and medication-related osteonecrosis of the jaws (MRONJ) are a growing problem. In 2016 a paper was published that reviewed the current literature regarding PRP and both preventing and treating ONJ. The study found several papers that used PRP as a prevention strategy, placing PRP in dental extraction sockets of high-risk patients and, as treatment, placing PRP after debridement of established osteonecrosis cases. The study stated there was inconclusive evidence to show benefit and randomized controlled trials were needed [44]. Some clinicals have reported PRP use after laser therapy in treating patients with MRONJ [45]. In 2018, a study was performed in rats which showed local application of autologous PRP was a viable therapy in preventing the occurrence of MRONJ following tooth extractions [46]. A systematic review in 2019 showed PRP as an adjuvant to surgical debridement can produce significant benefit in treatment of MRONJ, with one study finding 80.2% of patients completely healed [47].

TMJ

The use of PRP in temporomandibular joint disorders that are associated with chronic pain is gaining popularity as a treatment modality, especially as treating TMD can often be difficult [48]. A randomized controlled trial found injection of PRP in comparison to hyaluronic acid demonstrated more pain reduction [49]. A study was performed to determine the effects of PRP injections in cases of TMJ arthritis in domestic pigs. The authors found a "A significant reduction in signs of histological inflammation, such as hyperplasia of the synovial membrane, leucocyte infiltration, cartilage surface alterations, and an increase in cartilage-specific glycosaminoglycan content, was observed [50]."

Conclusion

Oral and Maxillofacial surgery covers a wide plethora of procedures, but dentoalveolar surgery is the most numerous in daily practice and therefore is likely why most available data on PRP and PRF pertains to extractions and dental implants.

PRP/PRF has potential to help with postoperative healing after extractions as well as improve bone and soft tissue formation for future dental implant site development. Investigations are starting to be done for use in TMJ disorders and bone reconstruction. PRP/PRF may also prove beneficial for some of the other procedures in the field, including soft tissue defects and flap reconstruction. As the reader can appreciate, the literature on this topic is lacking and therefore unable to provide definitive guidelines for its use. The safety and low cost of PRP/PRF along with the biological plausibility make it still a valuable treatment modality for practitioners. Over the next decade, the authors of this chapter expect to see a robust wave of literature of randomized controlled trials that will hopefully help develop meaningful clinical indications.

For now, the PRP/PRF literature appears centered around dental implant surgery. Dental implant osteointegration is already arbitrarily very successful. Implant success rate is estimated around 95% [14]. Most dental implant research focuses on how to improve success in less than ideal situations (i.e., inadequate bone height, uncontrolled diabetics). PRP and PRF may just be an additional tool in the dentists' arsenal to aid in these circumstances. While concrete evidence of actual osteointegration improvement with PRP/PRF is lacking, evidence does exist in terms of soft tissue healing and is likely to be a big area of future research. Soft tissue involving implants has been a long-studied area given the belief that it is directly related to the longevity of the implant's success. While the PRP/PRF may have clinical effects to primary or secondary stability, its use in promoting soft tissue healing cannot be ignored. PRP/PRF are minimally invasive, have essentially no risk to the patient, and can be inexpensive to produce. While some still believe there used to be use a better word here, the risk/benefit analysis with the current data supports its use in specific instances. Like much of dental research, without large-scale clinical trials, the data to fully support its use in everyday practice is limited.

References

1. Normando D. Third molars: to extract or not to extract? Dental Press J Orthod. 2015;20(4):17–8.
2. Branemark P, Zarb G, Albrektsson T. Tissue-integrated prostheses: ossesointegration in clinical dentistry. Chicago, IL: Quintessence Publishing; 1985.
3. Marx RE, Carlson ER, Eichstaedt RM, Schimmele SR, Strauss JE, Georgeff KR. Platelet-rich plasma: growth factor enhancement for bone grafts. Oral Surg Oral Med Oral Pathol Oral Radiol Endod. 1998;85(6):638–46.
4. Scully D, Naseem KM, Matsakas A. Platelet biology in regenerative medicine of skeletal muscle. Acta Physiol (Oxf). 2018;223(3):e13071.
5. Kobayashi E, Fluckiger L, Fujioka-Kobayashi M, Sawada K, Sculean A, Schaller B, Miron RJ. Comparative release of growth factors from PRP, PRF, and advanced-PRF. Clin Oral Investig. 2016;20(9):2353–60.
6. Ortolani E, Guerriero M, Coli A, Di Giannuario A, Minniti G, Polimeni A. Effect of PDGF, IGF-1 and PRP on the implant osseointegration. An histological and immunohistochemical study in rabbits. Ann Stomatol (Roma). 2014;5(2):66–8.
7. Strauss FJ, Stahli A, Gruber R. The use of platelet-rich fibrin to enhance the outcomes of implant therapy: a systematic review. Clin Oral Implants Res. 2018;29(Suppl 18):6–19.

8. Stahli A, Strauss FJ, Gruber R. The use of platelet-rich plasma to enhance the outcomes of implant therapy: a systematic review. Clin Oral Implants Res. 2018;29(Suppl 18):20–36.
9. Pachito DV, de Oliveira Cruz Latorraca C, Riera R. Efficacy of Platelet-Rich Plasma for non-transfusion use: overview of systematic reviews. Int J Clin Pract. 2019;73:e13402.
10. Al-Hamed FS, Tawfik MA, Abdelfadil E, Al-Saleh MAQ. Efficacy of platelet-rich fibrin after mandibular third molar extraction: a systematic review and meta-analysis. J Oral Maxillofac Surg. 2017;75(6):1124–35.
11. Bailey E, Kashbour W, Shah N, Worthington HV, Renton TF, Coulthard P. Surgical techniques for the removal of mandibular wisdom teeth. Cochrane Database Syst Rev. 2020;7(7): Cd004345.
12. Zhu J, Zhang S, Yuan X, He T, Liu H, Wang J, Xu B. Effect of platelet-rich fibrin on the control of alveolar osteitis, pain, trismus, soft tissue healing, and swelling following mandibular third molar surgery: an updated systematic review and meta-analysis. Int J Oral Maxillofac Surg. 2021;50(3):398–406.
13. Malhotra A, Kapur I, Das D, Sharma A, Gupta M, Kumar M. Comparative evaluation of bone regeneration with platelet-rich fibrin in mandibular third molar extraction socket: a randomized split-mouth study. Natl J Maxillofac Surg. 2020;11(2):241–7.
14. Moraschini V, Poubel LA, Ferreira VF, Barboza E. Evaluation of survival and success rates of dental implants reported in longitudinal studies with a follow-up period of at least 10 years: a systematic review. Int J Oral Maxillofac Surg. 2015;44(3):377–88.
15. Tabrizi R, Arabion H, Karagah T. Does platelet-rich fibrin increase the stability of implants in the posterior of the maxilla? A split-mouth randomized clinical trial. Int J Oral Maxillofac Surg. 2018;47(5):672–5.
16. Huang Y, Li Z, Van Dessel J, Salmon B, Huang B, Lambrichts I, Politis C, Jacobs R. Effect of platelet-rich plasma on peri-implant trabecular bone volume and architecture: a preclinical micro-CT study in beagle dogs. Clin Oral Implants Res. 2019;30(12):1190–9.
17. Diana C, Mohanty S, Chaudhary Z, Kumari S, Dabas J, Bodh R. Does platelet-rich fibrin have a role in osseointegration of immediate implants? A randomized, single-blind, controlled clinical trial. Int J Oral Maxillofac Surg. 2018;47(9):1178–88.
18. Lee JH, Nam J, Nam KW, Kim HJ, Yoo JJ. Pre-treatment of titanium alloy with platelet-rich plasma enhances human osteoblast responses. Tissue Eng Regen Med. 2016;13(4):335–42.
19. Yosif AM, Al-Hijazi A. Evaluation of the effect of Autolougues Platelet Rich Fibrin Matrix on osseointegration of the titanium implant immunohistochemical evaluation for PDGF-I&IGF-A. J Baghdad Coll Dent. 2013;25(1):70.
20. Temmerman A, Cleeren GJ, Castro AB, Teughels W, Quirynen M. L-PRF for increasing the width of keratinized mucosa around implants: a split-mouth, randomized, controlled pilot clinical trial. J Periodontal Res. 2018;53(5):793–800.
21. Taschieri S, Lolato A, Ofer M, Testori T, Francetti L, Del Fabbro M. Immediate post-extraction implants with or without pure platelet-rich plasma: a 5-year follow-up study. Oral Maxillofac Surg. 2017;21(2):147–57.
22. Kent JN, Block MS. Simultaneous maxillary sinus floor bone grafting and placement of hydroxylapatite-coated implants. J Oral Maxillofac Surg. 1989;47(3):238–42.
23. Schleier P, Bierfreund G, Schultze-Mosgau S, Moldenhauer F, Küpper H, Freilich M. Simultaneous dental implant placement and endoscope-guided internal sinus floor elevation: 2-year post-loading outcomes. Clin Oral Implants Res. 2008;19(11):1163–70.
24. Summer R. A new concept in maxillary implant surgery; the osteome technique. Compend Contin Educ Dent. 1994;2:152–62.
25. Aoki N, Kanayama T, Maeda M, Horii K, Miyamoto H, Wada K, Ojima Y, Tsuchimochi T, Shibuya Y. Sinus augmentation by platelet-rich fibrin alone: a report of two cases with histological examinations. Case Rep Dent. 2016;2016:2654645.
26. Lin YH, Yang YC, Wen SC, Wang HL. The influence of sinus membrane thickness upon membrane perforation during lateral window sinus augmentation. Clin Oral Implants Res. 2016;27(5):612–7.

27. Cabbar F, Guler N, Kurkcu M, Iseri U, Sencift K. The effect of bovine bone graft with or without platelet-rich plasma on maxillary sinus floor augmentation. J Oral Maxillofac Surg. 2011;69(10):2537–47.
28. Esposito M, Grusovin MG, Rees J, Karasoulos D, Felice P, Alissa R, Worthington H, Coulthard P. Effectiveness of sinus lift procedures for dental implant rehabilitation: a Cochrane systematic review. Eur J Oral Implantol. 2010;3(1):7–26.
29. Lekohlm U, Zarb G. Patient selection. Quintessence Publishing; 1985.
30. Benic GI, Bernasconi M, Jung RE, Hämmerle CH. Clinical and radiographic intra-subject comparison of implants placed with or without guided bone regeneration: 15-year results. J Clin Periodontol. 2017;44(3):315–25.
31. Benic GI, Hämmerle CH. Horizontal bone augmentation by means of guided bone regeneration. Periodontol 2000. 2014;66(1):13–40.
32. Sheikh Z, Hamdan N, Ikeda Y, Grynpas M, Ganss B, Glogauer M. Natural graft tissues and synthetic biomaterials for periodontal and alveolar bone reconstructive applications: a review. Biomater Res. 2017;21:9.
33. Wang HL, Misch C, Neiva RF. "Sandwich" bone augmentation technique: rationale and report of pilot cases. Int J Periodontics Restorative Dent. 2004;24(3):232–45.
34. Triplett RG, Schow SR. Autologous bone grafts and endosseous implants: complementary techniques. J Oral Maxillofac Surg. 1996;54(4):486–94.
35. Anitua E, Murias-Freijo A, Alkhraisat MH. Implant site under-preparation to compensate the remodeling of an autologous bone block graft. J Craniofac Surg. 2015;26(5):e374–7.
36. Levin L, Schwartz-Arad D, Nitzan D. Smoking as a risk factor for dental implants and implant-related surgery. Refuat Hapeh Vehashinayim (1993). 2005;22(2):37–43, 85.
37. Levin L, Nitzan D, Schwartz-Arad D. Success of dental implants placed in intraoral block bone grafts. J Periodontol. 2007;78(1):18–21.
38. Freilich M, Shafer D, Halepas S. Chapter 21: Dental implants. In: Ferneini E, Goupil M, editors. Evidence-based oral surgery. Springer Nature; 2019.
39. He L, Lin Y, Hu X, Zhang Y, Wu H. A comparative study of platelet-rich fibrin (PRF) and platelet-rich plasma (PRP) on the effect of proliferation and differentiation of rat osteoblasts in vitro. Oral Surg Oral Med Oral Pathol Oral Radiol Endod. 2009;108(5):707–13.
40. Borie E, Oliví DG, Orsi IA, Garlet K, Weber B, Beltrán V, Fuentes R. Platelet-rich fibrin application in dentistry: a literature review. Int J Clin Exp Med. 2015;8(5):7922–9.
41. Sakio R, Sakamoto Y, Ogata H, Sakamoto T, Ishii T, Kishi K. Effect of platelet-rich plasma on bone grafting of alveolar clefts. J Craniofac Surg. 2017;28(2):486–8.
42. Chen S, Liu B, Yin N, Wang Y, Li H. Assessment of bone formation after secondary alveolar bone grafting with and without platelet-rich plasma using computer-aided engineering techniques. J Craniofac Surg. 2020;31(2):549–52.
43. Marx RE, Armentano L, Olavarria A, Samaniego J. rhBMP-2/ACS grafts versus autogenous cancellous marrow grafts in large vertical defects of the maxilla: an unsponsored randomized open-label clinical trial. Int J Oral Maxillofac Implants. 2013;28(5):e243–51.
44. Lopez-Jornet P, Sanchez Perez A, Amaral Mendes R, Tobias A. Medication-related osteonecrosis of the jaw: is autologous platelet concentrate application effective for prevention and treatment? A systematic review. J Craniomaxillofac Surg. 2016;44(8):1067–72.
45. Fornaini C, Cella L, Oppici A, Parlatore A, Clini F, Fontana M, Lagori G, Merigo E. Laser and platelet-rich plasma to treat medication-related osteonecrosis of the jaws (MRONJ): a case report. Laser Ther. 2017;26(3):223–7.
46. Toro LF, de Mello-Neto JM, Santos F, Ferreira LC, Statkievicz C, Cintra L, Issa JPM, Dornelles RCM, de Almeida JM, Nagata MJH, Garcia VG, Theodoro LH, Casatti CA, Ervolino E. Application of autologous platelet-rich plasma on tooth extraction site prevents occurence of medication-related osteonecrosis of the jaws in rats. Sci Rep. 2019;9(1):22.
47. de Souza Tolentino E, de Castro TF, Michellon FC, Passoni ACC, Ortega LJA, Iwaki LCV, da Silva MC. Adjuvant therapies in the management of medication-related osteonecrosis of the jaws: systematic review. Head Neck. 2019;41(12):4209–28.

48. Anitua E, Fernández-de-Retana S, Alkhraisat MH. Platelet rich plasma in oral and maxillofacial surgery from the perspective of composition. Platelets. 2021;32(2):174–82.
49. Fernández-Ferro M, Fernández-Sanromán J, Blanco-Carrión A, Costas-López A, López-Betancourt A, Arenaz-Bua J, Stavaru Marinescu B. Comparison of intra-articular injection of plasma rich in growth factors versus hyaluronic acid following arthroscopy in the treatment of temporomandibular dysfunction: a randomised prospective study. J Cranio-Maxillofac Surg. 2017;45(4):449–54.
50. Naujokat H, Sengebusch A, Loger K, Möller B, Açil Y, Wiltfang J. Therapy of antigen-induced arthritis of the temporomandibular joint via platelet-rich plasma injections in domestic pigs. J Craniomaxillofac Surg. 2021;49(8):726–31.

PRP Dental Cases

9

Steven Halepas, Michael S. Forman, Xun Joy Chen, and Alia Koch

Case 1

A 55-year-old female presents with missing teeth at sites #19 and #30 with insufficient ridge (Seibert class I) for dental implant placement. The treatment options were discussed and the patient elected to undergo guided bone regeneration. Given this patient's defects, bilateral tunnel grafts were selected for augmentation.

Local anesthesia was delivered via bilateral inferior alveolar blocks and long buccal injections, utilizing 2% lidocaine with 1:100,000 epinephrine. At the start of the procedure, 20 cc of the patients' blood was drawn using a single-use 10 cc Red Top BD Vacutainer (R) Serum Blood Collection Tubes. The tubes were immediately placed in the centrifuge for 12 min at 2500 rpms (Fig. 9.1a). Figure 9.1b and c show the preoperative surgical sites and the horizontal ridge deficiency. A #15 blade was then used to make an incision in the buccal vestibule just lateral to the second premolars at the junction of attached and unattached tissue. A periosteal elevator was used to complete a subperiosteal dissection allowing for tunneling of the buccal aspect of the defective ridge (Fig. 9.1d). 1 cc of xenograft was placed in a 0.5 cc syringe with the top cut off (Fig. 9.1e). The bone was packed firmly into the buccal aspect of the defect (Fig. 9.1f). The platelet-rich fibrin (PRF) clot was then removed from the test tube using college pliers (Fig. 9.1g) and then flattened into a membrane and placed at the surgical sites bilaterally to aid in graft containment and as a biologic dressing (Fig. 9.1h). The surgical sites were then closed with 4-0 vicryl sutures (Fig. 9.1i). The tunnel graft procedure is an excellent technique that is

S. Halepas (✉) · M. S. Forman · A. Koch
Division of Oral and Maxillofacial Surgery, New York-Presbyterian/Columbia University Irving Medical Center, New York, NY, USA
e-mail: sh3808@cumc.columbia.edu

X. J. Chen
Will Surgical Arts, Rockville, MD, USA

© The Author(s), under exclusive license to Springer Nature Switzerland AG 2022
E. M. Ferneini et al. (eds.), *Platelet Rich Plasma in Medicine*,
https://doi.org/10.1007/978-3-030-94269-4_9

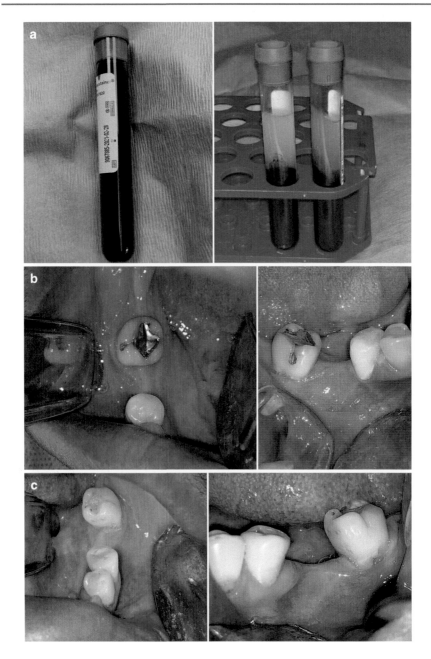

Fig. 9.1 (**a**) Two 10 cc tubes were filled with the patients' blood (left) and placed in the centrifuge for 12 min at 2500 rpms (right). (**b**) Site #30 preoperatively. (**c**) Site #19 preoperatively. (**d**) Surgical site with subperiosteal dissection. (**e**) Syringe loaded with bone particulate graft. (**f**) Bone particulate at intended site. (**g**) PRF being removed from tube and flattened into a membrane. (**h**) Platelet-rich fibrin membrane being placed over bone particulate at the incision site. (**i**) Post-surgical photo of the incision site with PRF membrane secured under the gingiva

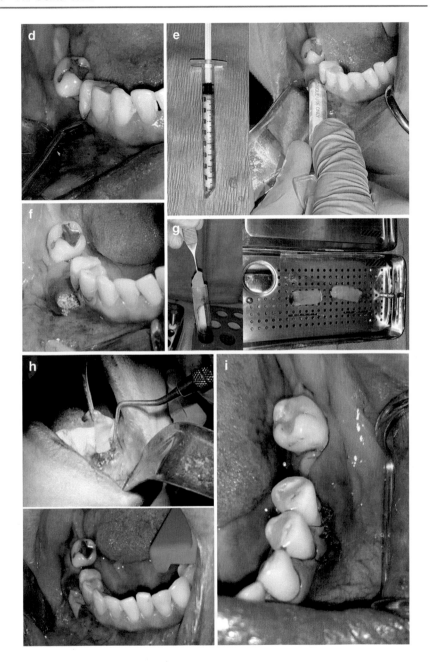

Fig. 9.1 (continued)

relatively minimally invasive with a small surgical incision. The gingival blood supply is not disturbed, and the use of a PRF membrane allows for nice soft tissue healing and minimizes bone particulate loss at the surgical site.

Case 2

A 48-year-old male presented with missing teeth #12 and 13. The patient elected for two dental implants at the edentulous sites. Due to pneumatization of the bone by the left maxillary sinus, a direct sinus lift was advised prior to proper implant placement. Local anesthesia was achieved with 4% Septocaine 1:100,000 epi via local infiltration. A #15 blade was used to make an envelope incision from tooth #11 to tooth #14 with a mesial release. A lateral sinus reamer bur was used to access the maxillary sinus. On exposure a tear was noted in the Schneiderian membrane (Fig. 9.2a). After discussing options with the patient, a PRF membrane was

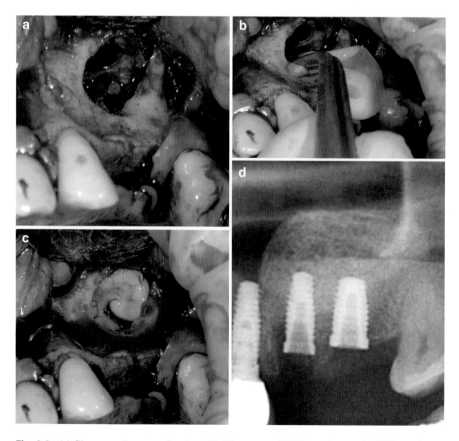

Fig. 9.2 (a) Sinus membrane perforation. (b) Placement of PRF membrane. (c) PRF membrane covering membrane defect. (d) Placement of membrane with simultaneous sinus lift

utilized. 10 cc of the patients' blood was drawn and placed in a centrifuge at 2500 rpm for 12 min. The PRF clots were flattened into a membrane and placed into the surgical site (Fig. 9.2b). The membrane was positioned over the defect, implants were subsequently placed and torqued to >35 N-cm, and the site was packed with particulate allograft (Fig. 9.2c, d). The tissue was reapproximated with 4-0 chromic gut sutures.

Case 3

A 42-year-old female presented with an atrophic mandible with what appeared as multiple traumatic bone cysts resulting in a defective ridge. Local anesthesia was achieved with 2% lidocaine 1:100,000 epi via inferior alveolar nerve block. A #15 blade was used to make an incision on the buccal aspect of the ridge with a mesial release. A full-thickness periosteal flap was raised, and a round bur was used to create bony windows into the cavity (Fig. 9.3a). Tissue was sent for biologic specimen. 10 ccs of the patient's blood were drawn into a test tube and placed in the centrifuge at 2500 rpms for 3.5 min. The top plasma/fibrin layer is removed with an empty syringe. This layer is the light orange layer noted in Fig. 9.3b. Generally, you can expect to obtain 1 cc of autologous fibrin glue for each 10 cc test tube of venous blood. Mix the autologous fibrin glue immediately with bone particles on a metal dish. Mold the bone into the desired thickness with a periosteal elevator. Leave the coated bone undisturbed on the metal dish for 5–10 min. The bone particulate will form "sticky bone" (PRF + allograft) seen in Fig. 9.3c. The sticky bone can then be placed at the surgical site with the desired morphology. A resorbable membrane was then placed over the sticky bone. The gingiva was closed with 4-0 chromic gut sutures. Figure 9.3e shows the cone beam computed tomography (CBCT) 5 months postoperatively.

Case 4

A 34-year-old male presented with a horizontal mandibular defect canine to canine. Treatment options were discussed and the patient elected to undergo a ramus graft procedure for anterior mandibular augmentation. Local anesthesia was achieved with 2% lidocaine 1:100,000 epi via inferior alveolar nerve block at the donor site and local infiltration at the surgical site. An incision at the donor site was performed on the buccal aspect of the ridge with a mesial release. A full-thickness flap is created to expose the mandibular body/ramus. A marking pen was utilized to mark the intended block size. A fissure surgical bur is used to make the superior cut along the external oblique and continued to create the anterior cut. The posterior and inferior cuts are scored with the bur without going to depth to avoid damage to the inferior alveolar nerve. A set of osteotomes are used to complete the cuts. A curved osteotome and mallet are used via the superior cut to complete the separation of the graft from the body of the mandible. The donor site was irrigated, and gingiva was closed with 4-0 chromic gut sutures. At the

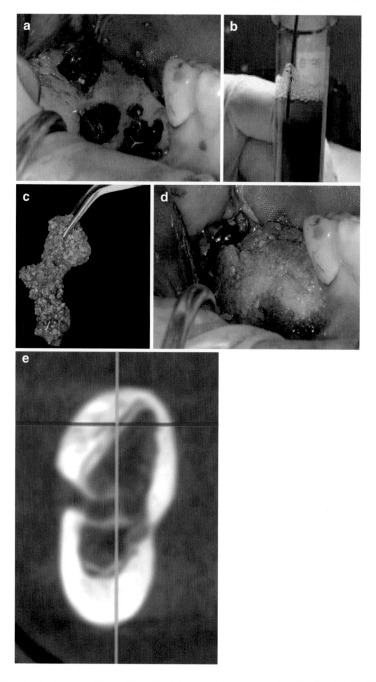

Fig. 9.3 (**a**) Atrophic mandible with multiple traumatic bone cysts. (**b**) Platelet-rich fibrin layer being removed from the tube to be mixed with the graft material. (**c**) "Sticky bone" consisting of the allograft and PRF. (**d**) Sticky bone placed at the surgical site in the desired morphology. (**e**) 5 months postoperative CBCT

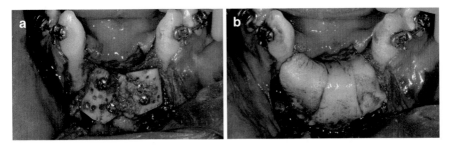

Fig. 9.4 (**a**) Ramus block grafts secured in place with titanium screws. (**b**) Three platelet-rich fibrin membranes covering the block grafts to aid in soft tissue healing

recipient site, a #15 blade was used to make an incision from canine to canine on the mandible with two vertical releasing incision with care to avoid damage to the mental nerve. A full-thickness flap was raised, and the block grafts were adapted and secured with titanium screws. Bone particulate was placed into defective areas. Perforation holes were drilled through the block grafts to encourage vascularization (Fig. 9.4a). 30 cc of the patient's blood was drawn and placed in the centrifuge for 12 min at 2500 rpms. Three PRF membranes were created and placed over the block grafts (Fig. 9.4b). The gingival flap was reapproximated and secured with 4-0 chromic gut sutures.

Case 5

A 61-year-old male presented following extraction of #9 by outside provider 3 months prior. The patient was interested in dental implant options. Due to the buccal defect, decision was made to undergo bone grafting augmentation. Local anesthesia was achieved with 4% Septocaine via local infiltration. A #15 blade was used to make an envelope incision with a distal release, and a full-thickness flap was raised exposing the defect (Fig. 9.5a). Particulate allograft was placed in the defective socket (Fig. 9.5b). A resorbable collagen membrane was placed over the bone particulate (Fig. 9.5c). 20 cc of the patient's blood was drawn and placed in the centrifuge for 12 min at 2500 rpms. Two PRF membranes were created and placed over the collagen membrane (Fig. 9.5d). The collagen membrane was used to add more support to the overlying PRF membranes. The PRF membranes were secured with 4-0 chromic gut sutures (Fig. 9.5e). The gingival flap was reapproximated and secured with 4-0 chromic gut sutures. Figure 9.5f demonstrates the 5-month postoperative CBCT showing good bone formation. At 6 months the patient underwent dental implant placement with good positioning noted in the periapical film in Fig. 9.5g.

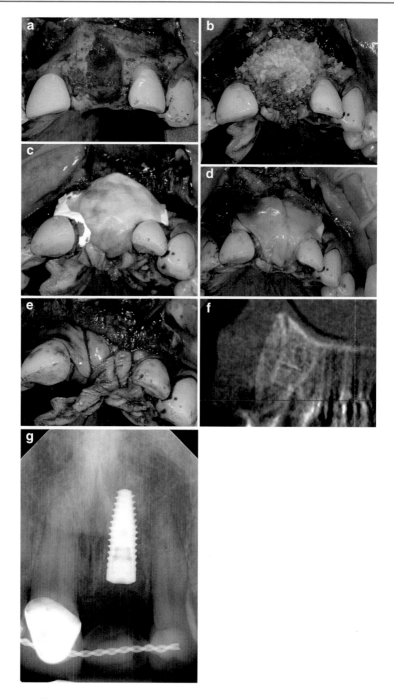

Fig. 9.5 (**a**) Exposure of the defect at previously extracted tooth #9. (**b**) Bone particulate placed in defect. (**c**) Collagen membrane placed over graft material. (**d**) Platelet-rich fibrin membranes placed over collagen membrane to aid in tissue healing. (**e**) Chromic gut suturing in netting fashion to retain membranes in place. (**f**) 5 month postoperative CBCT showing good bone remodeling in previous defect. (**g**) Periapical film after dental implant placement

PRP Complications

10

Andrew R. Emery and Elie M. Ferneini

Risks of Complications Among Types of Grafts

As with any drug or procedure in medicine or surgery, PRP use carries with it a risk of complications, albeit a low risk. The reason that PRP may produce very few complications is primarily because it is donated from the same patient to whom it is being administered (i.e., autologous graft) [1]. Conversely, cells and intercellular matrix that are not from oneself, such as from other humans (i.e., allogenic graft) or animals (i.e., xenogenic graft), carry with them foreign DNA unique to each donor. Consequently, this foreign DNA may be flagged by a recipient's immune system leading to various types of inflammatory responses. The end results of such detection are the activation of an inflammatory response leading to potential destruction of the transplanted graft and possibly also the recipients own tissues. To combat these inflammatory sequela, allogenic or xenogenic grafts can undergo treatment to destroy living cells leaving just the inert bony matrix behind [2]. If live allogenic cells must be transplanted, such as with stem cell transplants, the recipient is often given immunosuppressing medications to blunt their own immune response and decrease the risk of graft rejection. However, autologous PRP avoids these risks and thus at low risk of immune rejection and subsequent tissue damage. Nonetheless, despite the overwhelmingly benign nature of PRP, there exists a rather small cohort

A. R. Emery (✉)
Department of Oral and Maxillofacial Surgery, Massachusetts General Hospital/Harvard University, Boston, MA, USA
e-mail: aemery@partners.org

E. M. Ferneini
Beau Visage Med Spa and Greater Waterbury OMS, Cheshire, CT, USA

Department of Surgery, Frank H Netter MD School of Medicine, Quinnipiac University, North Haven, CT, USA

Division of Oral and Maxillofacial Surgery, University of Connecticut, Farmington, CT, USA

© The Author(s), under exclusive license to Springer Nature Switzerland AG 2022
E. M. Ferneini et al. (eds.), *Platelet Rich Plasma in Medicine*,
https://doi.org/10.1007/978-3-030-94269-4_10

of complications reported in the literature, which will be explored throughout this chapter.

As a point of clarification, when evaluating complications associated with PRP, it is important to attempt to separate the complications expected from the procedure itself (e.g., needle injection or incision) from those directly arising from PRP. Overwhelmingly, there is limited mention of complications related to PRP in the literature suggesting that it remains an area in need of additional research or attention to confirm its safety. By reviewing PRP complications throughout surgery and medicine, we can create a spectrum of adverse events, weigh the pros against the cons, and make well-informed clinical decisions with our patients.

PRP and Associated Complications

The number of studies reporting complications with PRP use is low. For example, a 2009 systematic review of 20 randomized control trials (RCT) of PRP use found only 6 reports of treatment-related adverse events, with just 2 of those studies identifying the type of adverse event [3]. Of these studies which did reported complications, the adverse outcomes were merely associated with PRP use, but there was not proof of actual causality. Nevertheless, the literature does describe various adverse events and special considerations that are worth mentioning, especially for providers using PRP to treat patients (see Table 10.1).

Complications Associated with Topical PRP Use

Topically applied PRP-like substances used for wound healing has been associated with few adverse events. One study compared the use of platelet gel, which is prepared from fresh autologous platelets and is similar to PRP, against saline for treating chronic venous leg ulcers [4]. The study reported two cases of dermatitis from the platelet gel group and two from the control group, suggesting comparable complications in the PRP and non-PRP groups. Additionally, the platelet gel group also reported a case of infection in an existing ulcer requiring a 10-day course of antibiotics. The platelet gel group also reported a case of thrombophlebitis associated with blood draws. Given the small sample size of the study (i.e., eight patients in the platelet gel group and seven in the control group), this single infection represents only weak evidence of an increased infection risk with platelet gels. However, it does advocate for vigilant observation for infection at the site of application of PRP-like products, and future research may attempt to elucidate if a relationship exists between topical PRP use and the risk of infection.

Complications Associated with Combining PRP with Surgery

A few studies have reported on the use of combining PRP with a surgical procedure. In select cases, PRP was found to increased surgical complication rates. For

Table 10.1 Complications associated with PRP and special considerations for use

Complications occurring more frequently in patients treated with PRP
Acromioclavicular osteoarthritis [6]
Hematoma [6]
Infection [4–7]
Nausea and dizziness [13]
Pain [6, 15–20]
Swelling [16–20]
Exanthematous itchy skin lesion [6]
Achilles tendon repair re-rupture [5]
Rotator cuff repair failure [6]
Shoulder stiffness [6]
Thrombophlebitis [4]
Complications occurring with equal frequently in patients treated with PRP
Leg compartment syndrome [7]
Post-dental extraction pain [8]
Intraoral swelling [8]
Trismus [8]
Oral infection [8]
Complications occurring less frequently in patients treated with PRP
Alveolar osteitis (aka dry socket) [10, 11]
Post-dental extraction pain [10, 11]
Blood urea nitrogen (BUN) levels [12]
Special considerations
Immunocompromised patients may not respond to PRP as well as immunocompetent patients [22, 23]
Use caution when harvesting autologous PRP from patients who have low hemoglobin levels [24]
Autogenous PRP is safer than allogeneic PRP for avoiding infectious disease transmission [21, 25, 26]
Bovine thrombin as an activator for PRP has been reported to cause immune-related factor V deficiency [27, 28]
Avoid PRP use near precancerous or cancerous lesions [21, 29]
Avoiding nonsteroidal anti-inflammatory drugs (NSAIDS) within 48 h to 1 week [21]
PRP may have dose-dependent effect on causing pain after injection into joints [18]

example, 1 study of 30 Achilles tendon repairs noted that 2 of the patients who were also treated with PRP developed adverse events including one deep space infection and one tendon re-rupture [5]. Further analysis was not pursued to determine if these complications were statistically significant between the groups. As such, it is unclear how much of a role PRP individually may play in such unfortunate outcomes compared to the procedure itself. Another study comparing PRP injections versus ropivacaine injections administered at the end of rotator cuff repair surgery found that PRP patients were 1.3 times more likely to experience a local adverse event compared to the ropivacaine group, but this was not statistically significant [6]. The local adverse events for both groups included shoulder stiffness and persisting or worsening pain, repair failure, infection, hematoma, acromioclavicular osteoarthritis, and an exanthematous itchy skin lesion. Similarly, another study reported superficial infections in 45% of injection group versus 30% of the control group, which was not statistically significant and was ultimately treated with oral antibiotics [7]. Given the lack of statistical significance, additional studies are needed to validate these claims that PRP is solely responsible for the aforementioned complications.

Furthermore, some surgical studies using PRP show no differences between the PRP and non-PRP groups. For example, a study of bilateral lower limb lengthening via tibia distraction with injections of bone marrow aspirate combined with PRP versus no injection found equal complication rates with one case of compartment syndrome per group [7]. Similarly, a prospective split-mouth study evaluating the efficacy of PRP after third molar removal found no differences in post-op pain, swelling, trismus, and infection compared to extractions sites that did not receive PRP [8]. In addition, another study found that platelet-rich fibrin (PRF), a growth factor similar to PRP, used with Bio-Oss® bone grafting material helps to promote bone formation in the femoral bones of dogs compared to Bio-Oss® alone, with no evidence of complications [9]. The comparable complication rates between PRP and non-PRP groups suggest that PRP does not elevate the risk of many post-surgical adverse events.

Lastly, some studies have also described decreased complications with PRP use including less pain and lower alveolar osteitis (aka dry socket) rates after tooth extraction when PRP was administered [10, 11]. PRP use has even been associated with a statistically significant reduction in blood urea nitrogen (BUN) levels compared to non-PRP groups [12]. Although these outcomes support the safety of PRP, the mechanisms leading to these results are unclear and call for future research for further elucidation.

Complications Associated with the Isolated Injection of PRP

A few studies have also focused on the complications associated with isolated PRP injections that are not associated with a surgical procedure. One study of providers injecting PRP into the knees of patients with osteoarthritis reported transient episodes of mild nausea and dizziness [13]. However, the physiologic explanation of such a response to PRP is unclear, and the authors of this chapter suggest these complications may be more likely from the blood draw or the painful stimulation of injecting into a joint. Conversely, other studies involving the intra-articular PRP have found no difference between the exposure and control groups. For example, a another study of 23 patients with patellar tendinopathy who were treated with either dry needling or injection of leukocyte-rich PRP found no differences in the complication rates [14]. The presence of weak or no evidence for PRP complications supports an overall high safety profile.

However, one of the most common side effects of injecting PRP is pain. One study compared PRP, glucocorticoids, and saline injections for 60 cases of lateral epicondylitis and found that 4 of the PRP patients had persistent pain days after injection [15]. Another study injected PRP into the knees of 60 patients with unilateral osteoarthritis and followed up 3 months later with serum Coll2-1 ELISA and found that PRP decreased collagen degradation products in the blood with few to no complications [16]. They noted mild knee pain and swelling, which they explained may be a result of mild inflammation from the PRP injection and subsequent distension of the synovial fluid by PRP fluid. Similarly, pain and swelling were also observed in other studies [17–20] following intra-articular PRP injections but with no other significant adverse events reported. Unfortunately, it is difficult to ascertain from these studies how much of the post-injection pain was attributable to the mechanical trauma of injecting a needle and fluid into a joint compared to the chemical or biologic properties PRP itself. Interestingly, the pain from PRP injection may be related to platelet concentration and dose dependent [18]. However, some physicians have argued that mitigating this post-injection pain with NSAIDS may be counterproductive as it can cause platelet inhibition [21]. For this reason, some clinicians advise avoiding NSAIDS within 48 h to 1 week of PRP injection [21].

Special Considerations

There are also unique scenarios when PRP use should be cautiously undertaken or even avoided.

Immunocompromised Patients

The unique therapeutic effect of PRP is related to the immune response that it triggers [22]. One study found that an inflammatory response to PRP was necessary to promote Achilles tendon repair in rats. In fact, pure PRP alone without inflammation was not effective at promoting tendon repair. These findings suggest that a patient's own immune system function or interindividual variability [23] may influence PRP effect and predict potential complications. For example, immunocompromised hosts or diabetics prone to poor wound healing may have decreased response to PRP therapy, which may affect the complications experienced. As a result, PRP is unlikely to be harmful in such cases but may have less effect or benefit.

Anemia

A potential relative contraindication to harvesting autologous PRP is a low hemoglobin level. Patients who are anemic or have low hemoglobin levels are inherently more sensitive to bloodletting of any amount so caution should be taken to avoid precipitating hemodynamic consequences resulting from the blood draw itself required to extract the PRP. Blood draws for PRP range between 20 and 60 ml [24], which is unlikely to worsen anemia in most cases but can in rare instances depending on one's blood volume and hemoglobin levels. A notable alternative at-risk patients may be allogeneic PRP, which unfortunately carries more potential risk of infection or immune rejection since it is donated from other patients [23]. Alternatively, if a procedure is elective and PRP is required, a provider may elect to optimize the hemoglobin levels in a patient before attempting the procedure at a later date.

Infection Transmission

The inherent, although low, risk of disease transmission from allografts and xeno-grafts has led some clinicians to pursue the even safer option of autografts [25]. A study from 2013 found no publications reporting on the risk PRP uniquely poses for infections, disease transmission (such as HIV, hepatitis, or Creutzfeldt-Jakob disease), immunogenic reactions, or any other adverse effects [26]. Similarly, an analysis of in vivo PRP studies from 1994 to 2019 found that PRP has a negligible risk of allergic reaction or transmission of blood borne microorganisms because of its autologous origin [21]. Therefore, the use of autologous PRP carries very little risk of infectious disease transmission and is naturally safer than allografts or xenografts.

Coagulopathy

Coagulopathy has also been associated with certain blood products. For example, there are reports of acquired immune-related factor V deficiency following topical application of bovine thrombin [27]. Interestingly, thrombin or calcium is often added to PRP to promote platelet degranulation and release of growth factors [28]. However, given the concern that topical bovine thrombin can rarely cause immune-related factor V deficiency, it may be best to allow PRP to naturally activate or to augment activation with calcium to avoid this risk completely. Currently, no evidence exists suggesting that pure autogenous PRP without additional activators causes any form of coagulopathy and thus may be the preferred protocol at this time to mitigate risk.

Precancerous and Cancerous Lesions

Some believe that PRP administration should be avoided in areas near malignant or dysplastic tissues because of the dense granules of growth factors potentially feeding carcinogenesis [21]. For the same reason, the intraoral application of PRP should be cautioned in patients with precancerous oral conditions and in areas of precancerous change such as oral leukoplakia, erythroplasia or solar cheilitis, and epithelial dysplasia [29]. Some studies also suggest avoiding PRP in any patient with a history of exposure to carcinogens (e.g., smokers, alcohol drinkers) or primary oral squamous cell carcinoma [29]. Despite the concern for PRP fueling a malignancy, others downplay these risks since PRP growth factors degrade in 7–10 days and cancer cells generally need more sustained growth factor exposure to thrive [26]. Overall, the literature is consistent in its recommendation to use PRP with caution in any anatomic locations concerning for premalignant or malignant processes.

Summary

Following a review of the current literature, PRP therapy results in very few complications overall. In fact, a lack of statistically significant differences between the complication rates of control and PRP groups in many studies suggests an overall acceptable safety profile comparable to non-PRP treatment options.

PRP is most often autologously derived and thus lacks foreign DNA that can incite immune responses seen with allogeneic or xenogeneic tissue grafts. However, the concentration of growth factors makes PRP risky in certain situations, like precancerous and cancerous tissues. Thus, a concerted effort to use PRP in appropriate candidates is required to minimize potential harm to patients.

As the field of medicine and surgery continue to report on the use of PRP, we hope that greater attention will be paid to looking for and acknowledging any complications that arise. Ultimately, such information will create a better understanding of the pathophysiology underlying PRP complications and can inform safe use practices to keep our patients healthy.

References

1. Benefits and associated risks of using allograft, autograft, and synthetic bone fusion material for patients and service providers – a systematic review. JBI Libr Syst Rev. 2010;8(8):1–12.
2. Tadjoedin ES, De Lange GL, Bronckers ALJJ, Lyaruu DM, Burger EH. Deproteinized cancellous bovine bone (Bio-Oss®) as bone substitute for sinus floor elevation. A retrospective, histomorphometrical study of five cases. J Clin Periodontol. 2003;30(3):261–70.
3. Martínez-Zapata MJ, Martí-Carvajal A, Solà I, Bolibar I, Ángel Expósito J, Rodriguez L, et al. Efficacy and safety of the use of autologous plasma rich in platelets for tissue regeneration: a systematic review. Transfusion. 2009;49(1):44–56.
4. Senet P, Bon FX, Benbunan M, Bussel A, Traineau R, Calvo F, et al. Randomized trial and local biological effect of autologous platelets used as adjuvant therapy for chronic venous leg ulcers. J Vasc Surg. 2003;38(6):1342–8.
5. Schepull T, Kvist J, Norrman H, Trinks M, Berlin G, Aspenberg P. Autologous platelets have no effect on the healing of human Achilles tendon ruptures: a randomized single-blind study. Am J Sports Med. 2011;39(1):38–47.
6. Flury M, Rickenbacher D, Schwyzer HK, Jung C, Schneider MM, Stahnke K, et al. Does pure platelet-rich plasma affect postoperative clinical outcomes after arthroscopic rotator cuff repair? Am J Sports Med. 2016;44(8):2136–46.
7. Lee DH, Ryu KJ, Kim JW, Kang KC, Choi YR. Bone marrow aspirate concentrate and platelet-rich plasma enhanced bone healing in distraction osteogenesis of the tibia. Clin Orthop Relat Res. 2014;472(12):3789–97.
8. Arenaz-Búa J, Luaces-Rey R, Sironvalle-Soliva S, Otero-Rico A, Charro-Huerga E, Patiño-Seijas B, et al. A comparative study of platelet-rich plasma, hydroxyapatite, demineralized bone matrix and autologous bone to promote bone regeneration after mandibular impacted third molar extraction. Med Oral Patol Oral Cir Bucal. 2010;15(3):483–9.

9. You JS, Kim SG, Oh JS, Kim JS. Effects of platelet-derived material (platelet-rich fibrin) on bone regeneration. Implant Dent. 2019;28(3):244–55.
10. Alissa R, Esposito M, Horner K, Oliver R. The influence of platelet-rich plasma on the healing of extraction sockets: an explorative randomised clinical trial. Eur J Oral Implantol. 2010;3(2):121–34.
11. Ogundipe OK, Ugboko VI, Owotade FJ. Can autologous platelet-rich plasma gel enhance healing after surgical extraction of mandibular third molars? J Oral Maxillofac Surg. 2011;69(9):2305–10. Available from: https://doi.org/10.1016/j.joms.2011.02.014.
12. Driver VR, Hanft J, Fylling CP, Berious JM. A prospective, randomized, controlled trial of autologous platelet-rich plasma gel for the treatment of diabetic foot ulcers. Ostomy Wound Manage. 2006;52(6):68–87. Available from: https://pubmed.ncbi.nlm.nih.gov/19543189/.
13. Patel S, Dhillon MS, Aggarwal S, Marwaha N, Jain A. Treatment with platelet-rich plasma is more effective than placebo for knee osteoarthritis: a prospective, double-blind, randomized trial. Am J Sports Med. 2013;41(2):356–64.
14. Dragoo JL, Wasterlain AS, Braun HJ, Nead KT. Platelet-rich plasma as a treatment for patellar tendinopathy: a double-blind, randomized controlled trial. Am J Sports Med. 2014;42(3):610–8.
15. Krogh TP, Fredberg U, Stengaard-Pedersen K, Christensen R, Jensen P, Ellingsen T. Treatment of lateral epicondylitis with platelet-rich plasma, glucocorticoid, or saline: a randomized, double-blind, placebo-controlled trial. Am J Sports Med. 2013;41(3):625–35.
16. Fawzy RM, Hashaad NI, Mansour AI. Decrease of serum biomarker of type II Collagen degradation (Coll2-1) by intra-articular injection of an autologous plasma-rich-platelet in patients with unilateral primary knee osteoarthritis. Eur J Rheumatol. 2017;4(2):93–7.
17. Kon E, Buda R, Filardo G, Di Martino A, Timoncini A, Cenacchi A, et al. Platelet-rich plasma: intra-articular knee injections produced favorable results on degenerative cartilage lesions. Knee Surg Sports Traumatol Arthrosc. 2010;18(4):472–9.
18. Filardo G, Kon E, Roffi A, Di Matteo B, Merli ML, Marcacci M. Platelet-rich plasma: why intra-articular? A systematic review of preclinical studies and clinical evidence on PRP for joint degeneration. Knee Surg Sports Traumatol Arthrosc. 2015;23(9):2459–74. Available from: https://doi.org/10.1007/s00167-013-2743-1.
19. Hassan AS, El-Shafey AM, Ahmed HS, Hamed MS. Effectiveness of the intra-articular injection of platelet rich plasma in the treatment of patients with primary knee osteoarthritis. Egypt Rheumatol. 2015;37(3):119–24. Available from: https://doi.org/10.1016/j.ejr.2014.11.004.
20. Jang SJ, Do Kim J, Cha SS. Platelet-rich plasma (PRP) injections as an effective treatment for early osteoarthritis. Eur J Orthop Surg Traumatol. 2013;23(5):573–80.
21. Mehrabani D, Seghatchian J, Acker JP. Platelet rich plasma in treatment of musculoskeletal pathologies. Transfus Apher Sci. 2019;58(6):102675. Available from: https://doi.org/10.1016/j.transci.2019.102675.
22. Dietrich F, Hammerman M, Blomgran P, Tätting L, Bampi VF, Silva JB, et al. Effect of platelet-rich plasma on rat Achilles tendon healing is related to microbiota. Acta Orthop. 2017;88(4):416–21.
23. Anitua E, Prado R, Orive G. Allogeneic platelet-rich plasma: at the dawn of an off-the-shelf therapy? Trends Biotechnol. 2017;35(2):91–3. Available from: https://doi.org/10.1016/j.tibtech.2016.11.001.
24. Dhurat R, Sukesh M. Principles and methods of preparation of platelet-rich plasma: a review and author's perspective. J Cutan Aesthet Surg. 2014;7(4):189.
25. Sánchez AR, Sheridan DDSPJ, Kupp MSLI. Is platelet-rich plasma the perfect enhancement factor? A current review. J Prosthet Dent. 2003;90(2):204.
26. Albanese A, Licata ME, Polizzi B, Campisi G. Platelet-rich plasma (PRP) in dental and oral surgery: from the wound healing to bone regeneration. Immun Ageing. 2013;10(1):1. Available from: Immunity & Ageing.

27. Foster KN, Kim H, Potter K, Matthews MR, Pressman M, Caruso DM. Acquired factor v deficiency associated with exposure to bovine thrombin in a burn patient. J Burn Care Res. 2010;31(2):353–60.
28. Cavallo C, Roffi A, Grigolo B, Mariani E, Pratelli L, Merli G, et al. Platelet-rich plasma: the choice of activation method affects the release of bioactive molecules. Biomed Res Int. 2016;2016:6591717.
29. Martínez-González JM, Cano Sánchez J, Gonzalo Lafuente JC, Campo Trapero J, Esparza Gómez GC, Seoane Lestón JM. Existen riesgos al utilizar los concentrados de Plasma Rico en Plaquetas (PRP) de uso ambulatorio? Med Oral. 2002;7(5):375–90.

Index

Printed by Printforce, the Netherlands